Texts in Philosophy
Volume 8

Bruno de Finetti
Radical Probabilist

Texts in Philosophy Series Editors
Vincent F. Hendricks   Vincent@ruc.dk
John Symons   jsymons@utep.edu

Bruno de Finetti
Radical Probabilist

Edited by
Maria Carla Galavotti

© Individual author and College Publications 2009. All rights reserved.
© Cover photograph Fulvia de Finetti

ISBN 978-1-904987-42-0

College Publications
Scientific Director: Dov Gabbay
Managing Director: Jane Spurr
Department of Computer Science
King's College London, Strand, London WC2R 2LS, UK

http://www.collegepublications.co.uk

Original cover design by Richard Fraser
Created by orchid creative   www.orchidcreative.co.uk
Printed by Lightning Source, Milton Keynes, UK

This book is dedicated to the memory of Dick Jeffrey (1926-2002) who would have been the natural candidate to deliver a paper in a workshop on Bruno de Finetti's radical probabilism had he lived long enough to be invited – in which case one might conjecture that he would have accepted with pleasure. His presence at the Bologna workshop was badly missed by participants, but the influence of his work is tangible in more than one of the papers belonging to this collection.

Contents

Contributors

GIORDANO BRUNO: Dipartimento di Metodi e Modelli Matematici per le Scienze Applicate, Università "La Sapienza", Roma, Italia.
bigi@dmmm.uniroma1.it

A. PHILIP DAWID: Centre for Mathematical Sciences. Wilberforce Rd., Cambridge, United Kingdom.
dawid@stats.ucl.ac.uk

FULVIA DE FINETTI: Piazza Filattiera 49, 00139 Roma, Italia.
fulvia.definetti@fastwebnet.it

MARIA CARLA GALAVOTTI: Dipartimento di Filosofia, Università di Bologna, Italia; Clare Hall, Cambridge, United Kingdom.
mariacarla.galavotti@unibo.it

DONALD GILLIES: Department of Science and Technology Studies, University College London, United Kingdom.
donald.gillies@ucl.ac.uk

COLIN HOWSON: Department of Philosophy, Logic and Scientific Method, London School of Economics and Political Science, United Kingdom.
c.howson@lse.ac.uk

GRAZIA IETTO-GILLIES: Centre for International Business Studies, London South Bank University, United Kingdom.
iettogg@lsbu.ac.uk

ALDO MONTESANO: Dipartimento di Economia, Università "Bocconi", Milano, Italia; Gonville and Caius College and Clare Hall, Cambridge, United Kingdom.
aldo.montesano@unibocconi.it

ALBERTO MURA: Dipartimento di Teorie e Ricerche dei Sistemi Culturali, Università di Sassari, Italia.
ammura@uniss.it

EUGENIO REGAZZINI: Dipartimento di Matematica, Università di Pavia, Italia.
eugenio.regazzini@unipv.it

ROBERTO SCAZZIERI: Dipartimento di Economia, Università di Bologna, Italia; Gonville and Caius College and Clare Hall, Cambridge, United Kingdom.
scazzieri@economia.unibo.it

BRIAN SKYRMS: Department of Logic and Philosophy of Science, University of California-Irvine, USA.
bskyrms@uci.edu

PATRICK SUPPES: Centre for Study of Language and Information, Stanford University, USA.
suppes@csli.stanford.edu

SANDY L. ZABELL: Department of Statistics, Northwestern University, USA.
zabell@math.northwestern.edu

Preface

MARIA CARLA GALAVOTTI

This volume stems from the workshop on "Bruno de Finetti, Radical Pro-babilist" held in Bologna on October 26-28, 2006 as part of the centenary celebrations commemorating de Finetti's birth. The idea behind the workshop was to focus not so much on de Finetti's main technical contri-butions to probability, but rather to draw attention to the philosophical aspects underlying his whole production. Reflecting the purpose of the workshop, the volume aims to shed new light on the multifarious perso-nality of Bruno de Finetti, who gave outstanding contributions not only to probability and statistics, but also to economics and philosophy. The arti-cles collected here therefore concentrate on the philosophical attitude Ri-chard Jeffrey labelled "radical probabilism"[1], imbuing all of de Finetti's writings. Special attention will be devoted to de Finetti's views on eco-nomics, which are inspired by the same philosophical approach, and an effort is made to highlight some lesser known aspects of his production.

The collection opens with a paper by Bruno's daughter Fulvia de Fi-netti, who recollects two important events in de Finetti's life, both of which occurred in Bologna. The first is the International Congress of Ma-thematicians that took place in 1928, where the young Bruno, aged 22 at the time, read his "Funzione caratteristica di un fenomeno aleatorio", con-taining the first formulation of exchangeability – the hallmark of de Fi-netti's approach to statistical inference. If the Bologna conference can

---

[1] See Jeffrey (1992a) and (1992b).

Bruno de Finetti, Radical Probabilist
Maria Carla Galavotti (ed.)
Copyright © 2009

ideally be regarded as the opening of de Finetti's career as a probabilist, the second event Fulvia mentions can be taken as its close: it amounts to the ceremony held two years before his death in which Bruno de Finetti was appointed Honorary President of the Italian Association for Mathematics Applied to Economic and Social Sciences, as a homage not only to his extraordinary contribution to probability and other fields, but also "for his dedication to Education and Culture", as reported by Fulvia.

In the second paper, Patrick Suppes reflects on various aspects of de Finetti's thought, and outlines some further developments. In the first place, he addresses de Finetti's pragmatism, that was inspired by the Italian philosopher Giovanni Vailati. There follows a discussion of de Finetti's position regarding the alternative between determinism and indeterminism, both of which were judged undecidable. Suppes proceeds to illustrate some results on the topic emerging from his own work and from that of other authors, that are very much in tune with de Finetti's conclusion. Another issue discussed in the paper is the axiomatization of qualitative probability, to which Suppes himself has given important contributions. Finally, the reader will find some intriguing remarks on the notion of coherence, that Suppes regards as more complex than is usually assumed by subjectivists.

Coherence is also the main focus of the article by Colin Howson, who first argues for an interpretation of coherence as logical consistency, and then fosters a view of the theory of probability as a logical calculus in which the finite version of additivity assumed by de Finetti could be embodied as an axiom. Although his proposal goes beyond de Finetti's production, Howson regards it as a continuation of de Finetti's work, sharing the same spirit.

Sandy Zabell addresses a topic that has received scant attention in the otherwise huge literature on de Finetti, namely his views on the meaning of probability in quantum mechanics. Zabell offers an in-depth analysis of various remarks on the topic to be found in de Finetti's writings from the early article "Probabilismo" (1931) to the Appendix to *Teoria delle Probabilità* (1970). His conclusion is that de Finetti's interest in quantum mechanics is motivated by its implications for the notion of event. By contrast, Zabell claims that de Finetti did not see any special problems for the subjective interpretation of probability arising from quantum mechanics.

Brian Skyrms devotes his paper to diachronic coherence, a topic to which he has given substantial contributions. Putting himself in the tradition inaugurated by Richard Jeffrey in a number of writings including *The Logic of Decision* (1965), Skyrms embraces a "minimal" approach, requiring minimal epistemological assumptions, leaving to particular contexts the burden of adding additional constraints, as required by the problems under study.

In their contribution, Philip Dawid and Maria Carla Galavotti argue that de Finetti's radical subjectivism should not be taken as a form of anarchism, or an "anything goes" approach according to which probability can take whatever value you like. De Finetti's well known claim that "Probability does not exist" is intended as a denial of objectivism, or the idea that probability depends entirely on some aspects of reality, not of objectivity. De Finetti took the problem of objectivity seriously, suggesting the adoption of scoring rules for an "empirical validation of probability assessments". The nature and role of this method is analysed by Philip Dawid in the second part of the paper.

Aldo Montesano's article calls attention to de Finetti's contribution to the problem of measuring risk aversion, an issue hitherto overlooked by the literature. Montesano shows that at the turn of the 1950s de Finetti found a measure of risk aversion that is a direct anticipation of the measure found in the 1960s by Kenneth Arrow and John Pratt. The short and somewhat elliptical presentation of this measure due to de Finetti is carefully reconstructed and compared with the Arrow-Pratt measure by Montesano who ends his article by pointing to new developments.

There follow two papers by Roberto Scazzieri and Grazia Ietto-Gillies focussing on de Finetti's contribution to economics. Scazzieri concentrates on de Finetti's theory of normative structures, which presupposes "awareness of context and ability to adjust partial objectives in the light of an open and pragmatic approach to the heterogeneity of goals". The implications of de Finetti's theory of normative structures for rational choice and economic analysis are discussed at length by Scazzieri, who draws attention to the illuminating distinction made by de Finetti between "situation" and "condition", which lies at the core of his analysis of the interplay of pure economics and policy objectives.

Grazia Ietto-Gillies, a former student of Bruno de Finetti, opens her paper with a vivid personal recollection of her Professor at the University of Rome, and proceeds to illustrate the main features of de Finetti's approach to economics. His criticism of the neo-classical theory is examined in some detail, as is his viewpoint, described as pluralistic and pragmatical, centred on individual behaviour, and deeply concerned with equity and social justice. It is precisely by virtue of the stress de Finetti placed on these issues that Ietto-Gillies regards his ideas on economics "more relevant than ever" today. This conclusion is also supported by the affinity between de Finetti's views and recent studies on happiness and society.

Eugenio Regazzini's paper examines some of Bruno de Finetti's contributions to statistics which have been neglected by the literature. In particular, Regazzini argues that de Finetti was able to prove a result on the "fundamental theorem of mathematical statistics" that is usually attributed to Francesco Cantelli. In addition, he presents some results on correlation and concordance found by de Finetti in the 1930s.

The paper by Alberto Mura deals with the logic of "trievents", namely the logic that develops a three-valued approach to the connective "|" used to represent conditional probability. After illustrating de Finetti's approach to the problem, as outlined in the paper read at the International Congress of Scientific Philosophy held in Paris in 1935, Mura makes a proposal for improving such approach.

The present volume ends with an Appendix on Bruno de Finetti's book *L'invenzione della verità* (The invention of truth) written in 1934 and published in 2006. In the first contribution to the Appendix, Fulvia de Finetti tells the story behind the book, namely she informs us that de Finetti, at that time lecturer of Infinitesimal Calculus at the University of Trieste, sent the text to the Royal Academy of Italy to take part in a competition for a grant, which in the end was awarded to someone else. In speculating on why de Finetti never published that text, Fulvia relates it to "Probabilismo", the most famous of de Finetti's philosophical papers, which appeared in 1931, recollecting the difficulties faced by Bruno in connection with its publication.

Among de Finetti's writings, "Probabilismo" is the one that comes closest to *L'invenzione della verità*, which contains an extensive presentation of the author's philosophical viewpoint. As Donald Gillies remarks in his paper for the Appendix, the text "is of great importance both as a philosophical work in its own right, and also for understanding the philosophical background to de Finetti's theory of probability". Gillies draws from de Finetti's text a vivid portrait of his philosophical perspective, which can be described as a blend of empiricism, pragmatism and operationalism.

In the paper that closes the Appendix Giordano Bruno puts forward an intriguing comparison between de Finetti's conception of knowledge, based on "certainty" rather than "truth", and the method adopted by Leonardo Sciascia in his book *La scomparsa di Majorana* (1975) in an attempt to find an explanation for the sudden disappearance of the physicist Ettore Majorana in 1938.

As a personal remark, let me add that *L'invenzione della verità* is a most interesting book that allows the reader to penetrate deep into de Finetti's philosophy that imbues his work on probability, statistics, economics, and his involvement with social and pedagogical matters. It should not pass unnoticed that the word "invention" in the title of the book points to the main trait of de Finetti's philosophical attitude, namely his "anti-realism"[2]. De Finetti uses the word "invention" to suggest that realism is an "illusion" and that all that belongs to our knowledge is the product of our thought, which in such an enterprise is guided not so much by "arbitrary laws", but rather by a "consideration of utility" (de Finetti 2006: 92).

The organizing committee of the workshop held in Bologna included Daniela Cocchi (Department of Statistical Sciences, University of Bologna), Roberto Scazzieri (Department of Economics, University of Bologna), Patrick Suppes (Emeritus, Stanford University) and Maria Carla Galavotti (Department of Philosophy, University of Bologna), who acted as the coordinator. In that capacity, and on behalf of the organizing committee, I wish to acknowledge the support of the "Bruno de Finetti Committee" created for the centenary celebrations commemorating de Finetti's birth; the University of Bologna - Alma Mater Studiorum; the Department of Philosophy, University of Bologna; the Department of Philosophy, University of Pisa; the Department of Statistical Sciences, University of Bologna; the Faculty of Humanities, University of Bologna; the "Cassa di Risparmio di Bologna" Foundation; the "Federigo Enriques" Interdisciplinary Research Centre for Epistemology and History of Science (CIRESS), University of Bologna; the "Dynamics of human knowldege" Focus Group of the Institute for Advanced Study (ISA) of the University of Bologna; the PRIN research group on "Probability, causal structures and intentional processes"; and the Research Group on "The identity of modern thought and its crisis" of the Department of Philosophy, University of Bologna.

---

[2] See Galavotti (1989).

References

de Finetti, B. (1931). "Probabilismo". *Logos*, pp. 163-219. Reprinted in *La logica dell'incerto*, ed. by M. Mondadori. Milan: Il Saggiatore, 1989, pp. 3-70. English edition "Probabilism". In *Erkenntnis* XXXI (1989), pp. 169-223.

de Finetti, B. (1992). *Probabilità e induzione (Induction and Probability)*, ed. by P. Monari and D. Cocchi. Bologna: CLUEB (special volume of *Statistica*, containing a collection of de Finetti's papers both in Italian and in English).

de Finetti, B. (2006). *L'invenzione della verità*. Milan: Cortina.

Galavotti, M. C. (1989). "Anti-realism in the Philosophy of Probability: Bruno de Finetti's Subjectivism". *Erkenntnis* XXXI, pp. 239-61.

Jeffrey, R. (1965). *The Logic of Decision*. Chicago: University of Chicago Press. 2nd edition 1983.

Jeffrey, R. (1992a). "Radical Probabilism (Prospectus for a User's Manual)". In *Rationality in Epistemology*, ed. by E. Villanueva. Atascadero, Cal.: Ridgeview, pp. 193-204.

Jeffrey, R. (1992b). "De Finetti's Radical Probabilism". In de Finetti (1992), pp. 263-275.

1

Some Links between Bruno de Finetti and the University of Bologna

FULVIA DE FINETTI

1 Introduction

The aim of this paper is to recollect two quite different events that took place at the University of Bologna, one at the beginning and one at the end of Bruno de Finetti's scientific career. In spite of the fact that he only spent a few days in Bologna, such events were very important for him, for different reasons.

The first is the International Congress of Mathematicians that was held in Bologna on September 3-10, 1928. It was there that de Finetti first made a name in the international academic world. He was next back in Bologna fifty-five years later to be awarded one of the very last tributes to his academic career.

In the meantime, the big masters of the Italian mathematical school had disappeared and dramatic changes in social and political life had taken place in Italy: years of war and hate, not yet completely overcome even today, among Italians who had to decide for which Italy to fight in World War II.

2 The 1928 International Congress of Mathematicians

I will start by explaining why that Congress was so important, not only for Bruno. At that time, I mean after World War I, the scientific world of mathematicians, or, if you prefer a more mathematical notation, the mathematicians set, was divided into two subsets: mathematicians be-

Bruno de Finetti, Radical Probabilist
Maria Carla Galavotti (ed.)
Copyright © 2009

longing to the winning countries and mathematicians belonging to the losing countries.

For this reason, previous congresses had failed to achieve a fully international standard. To his great credit, the prominent mathematician Salvatore Pincherle, at that time President of UMI (Unione Matematica Italiana) and also of the Mathematical International Union, succeeded in organizing a congress which, after many years of division, saw the participation of both winners and losers (Germans, Austrians, Bulgarians and Hungarians). In Germany there was a long discussion between the school of Berlin which was in favour of taking part in the conference, and that of Göttingen, which was against it. Out of a total of 836 participants, 76 were Germans: the biggest delegation, second only to the Italian one.

The University of Bologna, mother of all universities, would host the congress.

In order to encourage participation by the largest possible number of countries, invitations came directly from Giuseppe Albini, Rector of the University, who gave his opening speech in Latin: "*Vobis omnibus, viri clarissimi...*"

Then Pincherle, as president of the executive commission of the congress, gave his speech in French and traced the history of the congresses of mathematicians starting with the very first one, held in Zurich in 1897.

In the afternoon, the congress appointed Salvatore Pincherle as president, Ettore Bortolotti as secretary general and Leonida Tonelli as congress scientific secretary. Both these last had attended Pincherle's classes.

David Hilbert delivered the first address; the other speakers in the first general session were Jacques Hadamard and Umberto Puppini. Umberto Puppini was a professor of hydraulics at the University of Bologna; at that time he also played an important role in government, both local and central, and was a founder member of UMI.

On the following days general talks were given in the morning by Émile Borel, Oswald Veblen, Guido Castelnuovo, William Henry Young, Vito Volterra, Hermann Weyl, Theodor von Karman, Leonida Tonelli, Luigi Amoroso, Maurice Fréchet, Roberto Marcolongo, Nicolas Lusin and George Birkhoff. One may mention the title of Volterra's lecture: "La teoria dei funzionali applicata ai fenomeni ereditari" while Borel's, read by somebody else, was: "Le Calcul des probabilités et les sciences exactes".

Owing to the huge amount of talks submitted, more than four hundred, the congress divided into seven sections: *I Analysis, II Geometry, III Me-*

chanics, IV Actuarial, V Engineering, VI Elementary Mathematics, and VII History of Mathematics.

3 Bruno de Finetti

Who was de Finetti at the time? On November 21, 1927 he had graduated in mathematics *magna cum laude* at the University of Milan and by that time (1928), aged 22, he had already produced six articles: five, concerning Mendelian inheritance, had been published in *Metron, Rendiconti della Reale Accademia dei Lincei* and *Rivista di Biologia*. For the *Zentralblatt für Mathematik* Gumbel had reviewed "Conservazione e diffusione dei caratteri mendeliani. Caso panmittico" and "Conservazione e diffusione dei caratteri mendeliani. Caso generale", while "Probabilità che il massimo comune divisore di numeri scelti ad arbitrio sia un numero dato" appearing in *Rendiconti del Reale Istituto Lombardo di Scienze e Lettere* had been reviewed by Sternberg, again in the *Zentralblatt*. Both were German professors.

Bruno was working as head of the mathematical service at the Central Institute of Statistics in Rome. He was by no means completely unknown to the participants. Some of his professors at the University of Milan were there: Ugo Cassina who used "Latino sine flexione", the language devised by his master Giuseppe Peano, for his contribution, Umberto Cisotti elected president of one of the sessions, Bruno Finzi, Giulio Vivanti and Roberto Marcolongo who had given a talk in Milan on the new theories of Einstein while Bruno was a student there. He had also advised Bruno to accept the job at the Central Institute of Statistics.

Some professors from the University of Rome whom Bruno had met at various seminars he had attended, like Ugo Amaldi, Enrico Bompiani, Enrico Fermi, Giovanni Lampariello, Tullio Levi-Civita, Beniamino Segre, were also present. He was well acquainted with Guido Castelnuovo, who had often invited him to his home to check the progress of his work. It was to him that de Finetti sent the final version of his paper, and in a letter dated July 28, 1928 Castelnuovo recognized his ability as an analyst, gave advice on how to present it, and concluded: "*I feel sure that you will be able to make important contributions to Probability Calculus and its applications*". Luigi Galvani, who was his boss at the Central Institute of Statistics and Corrado Gini, President of the Institute, of course knew de Finetti very well.

4 More on the participants

The list of congress participants included three Nobel laureates: Niels Bohr, Nobel prize-winner for 1922, Enrico Fermi who would get it in 1938 and Max Born whose turn would come in 1954. However, none of them gave talks, and the *Proceedings* of the Congress offer no evidence of their participation in discussions.

The list also includes the names of Oskar Morgenstern (1902-1976) and Johann von Neumann (1903-1957), at that time freshly graduated and not yet John. Neither of them gave a paper, but there is evidence of their participation in the discussion in *Sections IV-B* and *I* respectively. I point this out because in 1969 my father wrote the "Preface" for the Italian edition of *Spieltheorie und Wirtschafttswissenschaft* by Oskar Morgenstern, published by Boringhieri, and in 1970 he wrote the chapter on "John von Neumann e Oskar Morgenstern" for the book *I maestri dell'economia moderna*, published by Franco Angeli. Two articles bearing the title "La teoria dei giochi" and "Riflessioni attuali sulla teoria dei giochi" appeared in the journal *Civiltà delle Macchine* in 1963.

A further name worth mentioning is that of George Pólya, who acted as president for one of the sessions in *Section Analysis*, where he presented a paper. As we will see later, he read another paper in the *Actuarial Section*. The reader might be interested to know that during a period spent in Rome around 1972, Pólya worked with Bruno on a documentary to be used to teach mathematics in schools, which featured an animated puppet representing a schoolboy called "Giorgetto" (Little George) after George Pólya. Let me add some more names: Jan Łukasiewicz (1878-1956), Karl Menger (1902-1985), Ernest Zermelo (1871-1953). Lukasiewicz gave his three presentations on Tuesday in the Section on Elementary Mathematics, with Alessandro Padoa, Adolf Abraham Fraenkel, Bernays, Alfred Tarski (1902-1983) and Cassina discussing it. Karl Menger presented his two talks on the same day in the Geometry Section. Zermelo gave no talk, but he contributed to the "interesting" discussion in the Section on Analysis regarding the papers by Adolf Abraham Fraenkel (1891-1965) and Stefan Mazurkiewicz (1888-1945). It may be said that the entire Gotha of the scientific world was there, and that no other congress of mathematicians, before or after, ever attained such a qualified audience.

5 Reading volume I of the Proceedings

Volume I, the first of the six volumes of *Proceedings*, contains a list of the four hundred papers delivered to the various sections, along with the general talks. Volumes II to VI contain the papers contributed. The names I quote in connection with the different sections correspond to those I found in the *Quaderno* (notebook) where Bruno used to record the names of people to whom he sent his own papers. It is quite interesting to follow address changes, as most of the time changes did not only concern the street or the city, but also the state and continent. The last sign is a cross indicating the person's death.

The section on *Analysis* was too crowded, which made it necessary to split it into four subsections; Giovanni Sansone, Waclaw Sierpinski, Octav Onicescu, Paul Lévy, Harald Bohr, Giuseppe Vitali, Renato Caccioppoli, Francesco Tricomi, Leonida Tonelli, Richard Courant, Luigi Fantappiè, Salvatore Pincherle, Giovanni Giorgi, and George Pólya contributed papers.

Berzolari opened *Section II* on *Geometry* and announced its division in two subsections. Gino Fano, Francesco Severi, A. Rosenblatt, Beniamino Segre, Enrico Bompiani, Guido Fubini, Alessandro Terracini, E. Bortolotti, Pia Nalli, and Ferdinand Gonseth gave papers.

The section on *Mechanics* divided in two subsections and George Birkhoff, Tullio Levi-Civita, Ferdinand Gonseth, Pietro Burgatti, Jacques Hadamard, Bohuslav Hostinsky, A. Rosenblatt, Mauro Picone and Giovanni Giorgi read papers. On Saturday the president of the session Giuseppe Gianfranceschi, Rector of the Università Gregoriana and President of the Pontificia Accademia dei Nuovi Lincei, gave a talk bearing the title "I criteri del calcolo delle probabilità dei fenomeni della radiazione", which unfortunately does not appear in the proceedings.

Bortolotti opened *Section V* devoted to *Engineering*. Bruno Finzi may be singled out for his contribution.

Ugo Amaldi opened *Section VI* devoted to *Elementary Mathematics* announcing the absence of Giuseppe Peano, caused by the death of his brother. One session in the following days was completely devoted to the teaching of Mathematics: Castelnuovo gave a full report of the meetings of the International Commission, established in 1908 during the International Congress of Mathematicians held in Rome. Some components of the Commission had died, among them Emanuel Czuber, the author of the book on *Probability*[1] that was the first on the subject that my father read. The same section hosted the contributions of Roberto Marcolongo and Ugo Cassina, who also read papers in *Section VII* on the History of Mathematics, where Ettore Bortolotti and Gino Loria also contributed papers.

I was surprised to find out that some of the participants gave more than one address, either in the same section or in different sections. One was Onicescu who read three papers, two in the Analysis and one in the Actuarial Section. Even more surprising was it to read that Hadamard, who delivered his talk in the first general session, also gave a paper in the Mechanics Section and took part in the discussion in other sections.

As for *Section IV Actuarial*, we are going to examine what happened in that section day by day, culminating on Saturday September 8, after a lunch offered by the Prefect of the Province of Bologna, when Bruno gave his talk "Funzione caratteristica di un fenomeno aleatorio", which polarised the attention of some big names in mathematics.

6   Section IV (Actuarial)

Corrado Gini opened the works on the Tuesday afternoon. After welcoming foreign participants, he announced that owing to the excessive number of papers submitted the section would include two subsections: *Section IV-A* devoted to *Mathematical Statistics and Probability*, and *Section IV-B* devoted to *Mathematical Economics and Actuarial Sciences*. He also drew attention to the contrast between the extraordinary number of papers this time and the few contributions presented at the Congress held in Rome twenty years earlier. He then called for strict cooperation between Mathematicians and Social Scientists. At the end of his speech, he explained the proposal by B. Lagunoff of the Statistical Office in Kiev to create an International Institute of Applied Mathematics. Many took part in the discussion, among them Toja, Du Pasquier, Cantelli, Hostinsky, Gumbel and Hadamard. Some of them advanced objections and the proposal was rejected. Then the presidents of the various sessions were appointed: for the one on *Mathematical Statistics and Probability* Francesco Cantelli, Gustav Du Pasquier, Eugene Slutsky and Maurice Fréchet

---

[1] See E. Czuber, *Wahrscheinlichkeitsrechnung und ihre Anwendung auf Fehlerausgleichnung, Statistik uns Lebensversicherung*, 2 voll., Leipzig: Teubner, 1903; 5th edition: 1938.

were nominated, while Luigi Galvani was appointed secretary for all four days.

Regarding *Subsection B Mathematical Economic and Actuarial Sciences*, let me just mention one of the presidents, namely Guido Toja, who was president on Tuesday when Gumbel and Ronald Aylmer Fisher gave their talks. Unfortunately, Fisher's presentation "The principle of synthetic induction in the estimate of subjective a priori probability" is not included in the Proceedings of the Congress. The *Proceedings* report that Fisher's contribution in *Section IV-A* was presented by session secretary Galvani. Again, there is no evidence that Fisher participated in the discussion, which tallies with the fact that, as far as I know, Bruno and Fisher had never corresponded and it was not until 1942 that Bruno started to send him his writings. An additional problem was due to language: Fisher's work became available in Italian in 1948, and my father only started to study English in 1950, before his tour of the U.S. sponsored by the Fulbright Foundation.

I mention Guido Toja because less than two years after the Bologna congress Bruno was awarded a prize bearing his name.

Let us now focus on *Section IV-A*, devoted to *Mathematical Statistics and Probability*. On Tuesday, with Cantelli acting as president, five papers were read and many contributed to the discussion. The talk by Dell'Agnola was followed by a number of comments by Cantelli, Slutsky, Fréchet, Khinchin, Castelnuovo, Neyman and Castellani, which were continued in one of the following sessions. Jerzy Neyman's talk had only one comment, by Molina.

On Wednesday Francesco Cantelli gave his presentation "Sui confini della probabilità", which prompted comments by Picone and Pólya; then Hostinsky gave his presentation, followed by Pólya, whose talk was commented upon by Neyman, Fréchet and Lévy. After some more talks, Jerzy Neyman pointed out the problem created by the huge number of papers on probability published in many journals, written in various languages, and the ensuing need to concentrate such publications in a restricted number of academic journals, so as to make it possible to keep pace with progress in this important field. Cantelli suggested postponing the discussion on this proposal until the following day, to allow time for exchanging ideas. The session closed at 7.30 p.m. According to the *Proceedings*, on the same day during *Subsection B* Cantelli read Brodie's paper, which gave rise to some discussion between Gumbel and Morgenstern.

The following day the appointed president Slutsky ceded the Presidency to Guldberg, who immediately opened the debate on Neyman's proposal. Cantelli suggested starting a Bulletin devoted to Probability, but Neyman proposed they adopt *Rendiconti del Circolo Matematico di Palermo* or *Metron*, both well known and widely circulating journals. Gumbel proposed a Commission be nominated to explore these possibilities and Jerzy Neyman, Carlo Alberto Dell'Agnola, Molina, Emil Julius Gumbel, and Wicksell were appointed as its members. Seven papers followed. As testified by the *Proceedings* Galvani, in his capacity as secretary presented Fisher's paper in his absence. Then Slutsky gave his presentation which was followed by a comment by Cantelli. The session ended at 8 p.m.

On Friday there were no sessions: three different excursions were organised for participants.

On Saturday, since Fréchet, who had been appointed president for the day, was engaged in another section, George Darmois became president. After the papers by Galvani and Gruzewski, Darmois invited Neyman to read the proposal by the Commission concerning publications on the Theory of Probability. By way of finding a solution to the dispersiveness of such publications, *Section IV-A* asked the Committee to publish an article in *Rendiconti del Circolo Matematico di Palermo* giving the statistics of publications appearing in different journals. These statistics corresponded to those mentioned in the last two volumes of *Jahrbuch Über die Fortschritte der Mathematik.*[2]

Section IV-A approved the proposal. The same commission also advanced the proposal to create an "International Committee for the progress of Probability calculus and its applications" with the aim of creating an International Society similar to "Circolo Matematico di Palermo" that would publish a Bulletin containing extensive summaries of all papers on the subject appearing in various languages. *Section IV-A* approved this proposal as well and the President invited those interested in being on the Committee to give their names and addresses.

---

[2] The *Jahrbuch Über die Fortschritte der Mathematik* (JFM) was founded in 1868 by the mathematicians Carl Ohrtmann and Felix Müller. In a preamble which appeared in the first issue of the journal, the editors described the aim of this service as follows: "Our task is twofold: firstly, we intend to offer those who are unable to trace all publications appearing in the field of Mathematics a tool enabling them to follow the evolution of mathematical sciences; secondly, we want to help active scientists to find out about the results obtained in the field of Mathematics".

Twenty-one names and addresses follow. Since the order in which they are listed is neither alphabetic nor by country, I presume it reflects the order in which people volunteered:

F.M. Urban (Brno, Czechoslovakia)
Octav Onicescu (Bucharest)
Emil Julius Gumbel (Heidelberg)
Carlo Alberto Dell'Agnola (R. Istituto Superiore di Scienze economiche e commerciali, Venice)
Bruno de Finetti (Istituto Centrale di Statistica, Rome)
S.D. Wicksell (University, Lund, Sweden)
E. C. Molina (195 Broadway, New York)
George Darmois (8, Rue du Hout Bourgeois, Nancy)
Eugene Slutsky (Nikitsky Boulevard 12/18, Moscow, 19)
Francesco Paolo Cantelli (via Merulana 105, Rome)
Aleksandr Yakovlevich Khinchin (I Université, Institut Mathématique, Moscow)
L. Gustave Du Pasquier (Sablons, 88 Neuchatel, Switzerland)
Jerzy Neyman (Institut Mathématique de l'Université N. Swint, 72, Warsaw)
Antoni Lomnicki (Politechniki, Lwòw, Poland)
Gruzewski (École Sup. de Commerce, Warsaw)
Corrado Gini (Istituto di Statistica e Politica Economica R. Università, via delle Terme di Diocleziano, 16, Rome)
C. Jordan (28 Szerbutca, Budapest)
G. Pietra (R. Università, Padua)
K.G. Hagstrom (Livfovrsakringsbolaget, Frambinden, Stockholm)
Y. Miura (Chuo, Central University of Tokio, Kanda)
V. Korinek (Haute École Technique Karlownam, Prague II).

Fréchet resumed the chair to let Darmois give his contribution, which was followed by comments from Cantelli, Fréchet and others. The following contribution again brought comments by Cantelli and Gumbel. At that point another change of presidency took place: Fréchet stood down and Gini acted as President in his place. The following day people were supposed to leave Bologna, as the closing session was scheduled to take place in Florence. This atmosphere of winding down surrounded the last four presentations, the last but one being that by de Finetti. There were no comments on the last four papers, and the session closed at 7.45 p.m. This concludes our report of the very last session.

This is the chronicle of *Section IV-A*. I believe that what happened that day, namely the creation of an "International Committee for the progress of Probability Calculus and its applications", was really important for the emergence of Probability as an autonomous scientific discipline. People who signed up regarded Probability as an important field of research and one that marked the birth of a new branch of Mathematics. Though it is always difficult to establish a starting date and obviously the subject was not new, there can be little doubt that from this point on it acquired a new dignity of its own. For this reason I would pick September 8th, 1928 as the key date. Of course I recognize that mine is a subjective point of view which may have something to do with DNA.

Let me briefly comment on the list. Fourteen of the names also appear in the second part of the *Quaderno* (notebook) that I have already mentioned.

Most people on the list had presented a paper in that subsection. The absence of Fisher is puzzling but there again he was not present even for his paper in *section IV-A*. It might be wondered why he did not present his contribution personally. One might conjecture some kind of mismatch between Fisher and Neyman. Undoubtedly, Neyman's personality, full of ideas and energy, comes out very clearly, even in the concise summary in the *Proceedings*. So maybe Fisher did not like to see the leading role played by the younger Neyman. In fact, as we will see in a moment, Fisher also skipped the meeting in Geneva in 1937.

I will add some information about those on the list who shared some biographical traits with de Finetti, or with his interests, and when appropriate give some details as to their relationship with him, if only to give an idea of the personalities he had to measure himself against.

Corrado Gini (1884 - 1965)

Italian statistician and economist, who studied law but also attended classes in mathematics. He became Director of ISTAT in 1926 and in 1927 created the "Istituto Centrale di Statistica", becoming its first President. He enormously influenced the development of statistics in Italy, from both a scientific and political perspective, and for a long time represented the Italian statistical school.

He exercised an enormous influence on my father's life and career: in 1926 *Metron* – the journal he had founded in 1920 – published my father's very first article, and Gini offered him a job as soon as he finished his studies. At that time my father was still a student at the Polytechnic and the offer of employment convinced my grandmother to let Bruno move to the Mathematics Faculty of the University of Milan. Once in Rome, he had the opportunity to meet a number of important professors that worked there, such as Guido Castelnuovo, Federigo Enriques, Enrico Fermi, Mauro Picone, Octav Onicescu, to name only a few. Since he worked nearby, after work he used to join the seminars that took place in Panisperna street, and was particularly impressed by Enrico Fermi, whose rapid career became a target for himself. Rome was also where he met Renata, his future wife and my future mother. A big influence indeed!

Gini is likely to have also influenced my father's decision to leave the Istituto Centrale di Statistica and join Assicurazioni Generali in Trieste. The letter from my father to Assicurazioni Generali is dated 1929 and in 1929 Gini published *Di una applicazione del metodo rappresentativo all'ultimo censimento italiano della popolazione* in *Annali di Statistica*. In a letter dated April 15, 1929, Bruno wrote to his mother that the volume of *Annali di Statistica* contained the first work he had done for the Institute and particularly the two Appendices, the most difficult part of the work, were completely his own, but the names were only those of Gini and Galvani! For this or some other reason Bruno left the Institute at the end of his three year contract in 1931.

Francesco Paolo Cantelli (1875 - 1966)

Born in Palermo, he completed his studies in mathematics at the University of Palermo in 1899. He first worked in astronomy and through the information to be found in the *Divina Commedia* he showed that the imaginary journey of Dante took place in 1301. Then for twenty years he worked as an actuary at the Istituto di Previdenza della Cassa Depositi e Prestiti. In 1923, he obtained the chair of Financial and actuarial mathematics in Catania. In 1931 he became a professor in Rome and kept that post until retirement. He died in Rome. He made contributions to probability as well as financial and actuarial mathematics. In 1929 he founded the *Giornale dell'Istituto Italiano degli Attuari* which he directed until 1958. (In this journal Bruno published several works.) He was elected member of the Accademia dei Lincei in 1947.

The differing views on probability held by Cantelli and Bruno caused my father some trouble at the time he decided to move to the University of Rome, and again in connection with his election to the Accademia dei Lincei. My father did not become a member of the Accademia until 1974.

Octav Onicescu (1892 - 1983)

Mathematician, founder of the Romanian school of probability theory, founder of the school of statistics. Octav Onicescu was the first Romanian who obtained a doctorate in mathematics in Rome with the qualification *magna cum laude*, in front of a commission of eleven university professors, chaired by Levi-Civita.

He received two Rockefeller fellowships, the first in 1924 and the second in 1933.

Bruno and Onicescu first met in Rome and later saw each other in Geneva in 1937. Their long friendly acquaintance gave my father the opportunity to publish one of his many works on probability – "Sul ruolo concettuale delle considerazioni asintotiche nella teoria delle probabilità" (written in 1968) – in the book *Calcolo delle probabilità e applicazioni* that Onicescu published in Italian in 1969, and also to contribute two short papers to the book *Papers in honour of Octav Onicescu* on the occasion of his 90th birthday in 1982.

I myself met him and his wife on two occasions, the first in Rome and the second in Bucharest in September 1971 for the Congress on Logic, Methodology and Philosophy of Science.

That Congress was also the last chance for my father to meet his dearest friend Jimmie Savage, who was there as an invited lecturer. The title of his lecture was *Probability in Science: A Personalistic Account*. The sudden news of Savage's death came as a shock for my father, who lost the only person able to understand and share his viewpoint on probability, and ended a fruitful twenty years long correspondence. I could add much about their close affinities, but that would take us far from Bologna!

Emil Julius Gumbel (1891 - 1966)

German statistician, born in Munich. Like many other Europeans, he migrated to the United States in 1940 for political reasons. He met Bruno in Geneva in 1937 for the Colloquium on probability. He died in New York City. He reviewed some works by de Finetti for the *Zentralblatt*.

George Darmois (1888 - 1960)

In April 1929 he wrote a letter to my father to thank him for sending his work "Sulle probabilità numerabili e geometriche" and to inform him that in March he had discussed Bruno's papers on Mendelian characters in his seminar. He also mentioned that Hadamard had told him that Bruno would be in Paris the following year, and that he had attended Fermi's lectures at the Institute Henri Poincaré, adding a very kind comment on the vitality of Italian scientific research.

Carlo Alberto Dell'Agnola (1871 - 1956)

In 1926 he became professor of Matematica Finanziaria at the Faculty of Economia e Commercio in Venice (Cà Foscari).

Aleksandr Yakovlevich Khinchin (1894 - 1959)

Khinchin's father was an engineer. Khinchin attended the technical high school in Moscow where he became fascinated by mathematics. However, mathematics was certainly not his only interest when he was at secondary school, for he also had a passion for poetry and theatre. He completed his secondary education in 1911 and entered the Faculty of Physics and Mathematics of Moscow University in that year.

When attending the university in Moscow Khinchin was an outstanding student; he wrote his first paper in 1916, before graduating. After graduating in 1916, Khinchin remained at Moscow University doing research for his dissertation which would enable him to become a university teacher.

Around 1922 he took up a new mathematical interest when he began to study probability. In 1927, Khinchin became a professor at Moscow University and in the same year published *Basic Laws of Probability Theory*. He reviewed de Finetti's "Le leggi differenziali e la rinuncia al determinismo" for the *Zentralblatt*.

Antoni Lomnicki (1881 - 1941)

Polish mathematician who became a professor in 1920. He was killed by the Nazis during the Second World War.

Charles Jordan (1871 - 1959)

He pioneered probability theory and statistics in Hungary.

Eugene Slutsky (1880 - 1948)

Eugene (or Eugen or Yevgeni) Slutsky wanted to become a mathematician, but was expelled from the University of Kiev for participating in student revolts. He eventually became a doctor of Law and in 1911 started to teach at the Kiev Institute of Commerce. He published his first economics paper in 1915 in the Italian *Giornale degli Economisti*. His other prominent contribution to economics came in 1927. The rest of his work, accomplished for the most part after he moved to Moscow in 1920, was in probability theory.

Gaetano Pietra (1879-1961)

Professor of Statistics at the University of Padua. He was a Senator for the Christian Democrat party in 1948-53.

Jerzy Neyman (1894 - 1981)

I will spend a few more words on Jerzy Neyman's profile for two reasons. The first is that his was one of the three names — the others being those of Castelnuovo and Frechét — that my father mentioned in his well-known Farewell Lesson, in addition to that of Jimmy Savage. On important occasions these mathematicians gave de Finetti the opportunity to put forward his ideas, even though these were in contrast with their own. That is something that Bruno appreciated a lot. The second reason lies with some similarities in their lives.

Jerzy was born of Roman Catholic parents who considered themselves Polish and were Polish-speaking, although at the time Poland did not officially exist as a separate country. For a few years Neyman wrote papers under the name of Splawa-Neyman, the first part being a mark of nobility.

As a young boy, Jerzy lived in several different towns: up to the age of ten a governess taught him at home and then he entered the local gymnasium. Remarkably, by that time he could speak five languages: Polish, Ukrainian, Russian, French and German. In 1906 his father died and his mother, who was left with little money to bring up her son, moved to Kharkov where she had relatives. Neyman excelled at the gymnasium in Kharkov and decided to study mathematics at university. After completing high school and before enrolling at university he went on a European tour by train through Austria and Italy.

Neyman began his studies at Kharkov University in the autumn of 1912. He was interested both in physics and in mathematics. When World War I started in 1914, many students left for military service, but Neyman failed the eyesight test and remained at the University. He wrote a paper in 1915. During the academic year 1915-16 he attended a class on probability given by Aleksandr Bernstein, and was strongly influenced by him.

In September 1917, having completed his curriculum, Neyman remained at Kharkov University preparing for an academic career and began to take an interest in statistical ideas. However, the last year of the war, the Russian Revolution, and the civil war, totally disrupted academic life of the University. Neyman's health began to deteriorate. The doctors diagnosed tuberculosis. Poland and Russia were at war and Neyman was imprisoned and held for about six weeks. Despite the difficulties he had to undergo, Neyman passed his examinations and became a lecturer at Kharkov University. In 1921, he went to Poland and made contact with Sierpinski. To earn some money he took a job as a senior statistical assistant at the Agricultural Institute in Bromberg. Keen to be in Warsaw, in December 1922 he took a job in the State Meteorological Institute there. He began to write papers on the applications of statistics. He became an assistant at Warsaw University at the beginning of the academic year 1923-24 and obtained a doctorate in 1924 with a dissertation on the application of probability to agricultural experimentation. After being awarded a Rockefeller Fellowship to work in London, he moved there in September 1925. He then obtained an extension of that fellowship which allowed him to spend one year in Paris. He arrived in Paris in the summer of 1926, and went to visit Borel.

Neyman returned to Poland in May 1927 and immediately tried to set up a biometric laboratory in Warsaw. He spent time in both Warsaw and Krakow. On June 26 he obtained his *habilitation* and began lecturing as a *dozent*. In 1928 he set up a Biometric Laboratory at the Institute for Experimental Biology in Warsaw. In 1934 he obtained a three-month leave of absence to go to England, and extended it until 1938. Although Fisher had inspired much of Neyman's work, once they were working in the same building relations seemed to break down. In the spring of 1937 Neyman spent six weeks in the United States on a lecture tour. Shortly after his return he was offered a lectureship at the University of California at Berkeley.

In 1937 Neyman was in Geneva at the famous Colloquium, where some of the participants in the Bologna congress, namely Fréchet, Hostinsky, Lévy, Onicescu and Pólya, had a new opportunity to discuss their different viewpoints about probability. There he saw Bruno again.

In April 1938, Neyman accepted the offer from Berkeley and worked there for the rest of his life, teaching probability and statistics in the Mathematics Department. After 1945, Neyman, who was director of the Statistical Laboratory, organized the Berkeley Symposia on Mathematical Statistics and Probability. These took place at five-year intervals and brought together for a period of six weeks a large number of eminent scholars. In 1950, Bruno visited the United States for three months; for the occasion he studied English and attended the second Berkley Symposium. Neyman received him with tokens of great friendship and it was on that occasion that he met Leonard Jimmie Savage, who immediately invited him to Chicago. In 1951, Neyman wrote to de Finetti that he was surprised not to find the latter's name among the members of the International Statistical Institute. He promoted Bruno's membership and said he would start the procedure, asking him to find one Italian member willing to support it, because usually at least one of the backers belonged to the country of the candidate. This is how Bruno became a member of ISI. In 1957, he also became a member of the Istituto Centrale di Statistica.

In his note "La scienza e la lotta contro l'incertezza", published in 1962, Bruno invites the reader to read Neyman's "Indeterminism in science and new demands on statisticians", referring to it as a "masterly exposition". In the note immediately following this one, he mentions two works by R.A. Fisher on the subject of how to design and analyze experiments, a subject to which, according to de Finetti's note, "Fisher gave the greatest impulse, despite some conceptual imperfection partly corrected by Neyman and Egon Pearson and partly by the subjectivist school...".

Let me add a few words on the already-mentioned conference that took place in Geneva in 1937 – the famous Colloquium on Probability, which may be considered as the first meeting of the Committee. Among the participants we find de Finetti, Gumbel, Neyman, Onicescu, whose names appeared in the list, and in addition Paul Lévy, Heinz Hopf, Wolfgang Doeblin and Bohuslav Hostinsky, George Pólya and Maurice Fréchet, who were also in Bologna; but many new entries in the field were also there. I should mention by the way that the merit of having produced a very well-documented discussion of the various viewpoints that were advanced on probability during that conference goes to my father.[3]

7   Some consequences of the Congress

I will now try and recollect the many opportunities that came Bruno's way from his participation in the Bologna congress. The work presented by de Finetti at that Congress was published with some additions in *Rendiconti della Reale Accademia dei Lincei* sponsored by Castelnuovo and Levi-Civita. This was the first of a series of papers by de Finetti that Castelnuovo presented at the Accademia dei Lincei, the last of which was read in 1933, when de Finetti had already left Rome. Castelnuovo was also president of the commission for the Toja award won by de Finetti in 1930 and a member of the commission for the Milan Insurance award obtained by de Finetti in 1934 – the president of this commission was Tullio Levi-Civita. Three months after the Bologna congress, Bruno became a member of the "Circolo Matematico di Palermo", the international institution founded in 1884.

In 1929 and 1930, Giovanni Giorgi presented seven notes by de Finetti at the "Pontificia Accademia delle Scienze Nuovi Lincei".

Moreover, Bruno surely captured the attention of Sierpinski, one of the founders of the famous journal *Fundamenta Mathematicae* which in 1931 published Bruno's article "Sul significato soggettivo della probabilità". It is worth noticing that Sierpinski was also a member of the editorial board of *Rendiconti del Circolo Matematico di Palermo*.

Let me translate a passage from a letter that Bruno wrote to his mother a few months after the conference, which testifies to the importance for him of having come to be known through his participation in that congress: "Today I received a letter from Vivanti informing me that he had received a very flattering statement from Hadamard concerning myself … Would you believe that Hadamard knew my works on Mendelian laws perfectly and that he gave them to Professor Darmois to be examined in his Mathematical Seminar in Paris … Then Vivanti transcribed for me this passage from Hadamard that I copy for you: C'est vous dire que je suis tout convaincu de sa valeur. Je serai très heureux de le voir avec nous a Paris". These words refer to a Rockefeller fellowship that Bruno would

---

[3] See B. de Finetti, "Resoconto critico del colloquio di Ginevra intorno alla teoria delle probabilità", in *Giornale dell'Istituto Italiano degli Attuari* 9, 1938, pp. 1-40. Also "Compte rendu critique du colloque de Genève sur la théorie des probabilités", in *Actualités Scientifiques et Industrielles* 766 (VIII), Paris: Hermann, 1939.

have liked to get while he was a student, but Vivanti suggested that it would be easier to get it after graduation. Immediately after graduation he accordingly started the necessary procedures. The difficulty at that point came in connection with his job, which he could keep at most for an absence of six months. Moreover, they judged him to be too young. It was only in May 1935 that Bruno went to Paris, upon an invitation to give a series of conferences at the Institute Poincaré.

After this long excursus on the 1928 Congress of Mathematicians, I come briefly to the second event that brought Bruno to Bologna, which was somehow a consequence of the first.

8   The honorary presidency of AMASES

On the 5th of November, 1983 the University of Bologna hosted some one hundred and fifty professors from various universities for the ceremony organized by AMASES, the Italian Association for Mathematics Applied to Economic and Social Sciences, jointly with a Committee especially nominated to honour de Finetti as Honorary President of the Association.

The ceremony took place in the old rectory that had hosted the congress where Bruno delivered his important contribution, the first he had ever given at an International Mathematics Conference. Professor Rizzoli, the Rector of the University, in his opening speech explained the reason for this choice. As one may easily imagine, the official speakers had words of appreciation for de Finetti's scientific merits and for his dedication to Education and Culture, which I would rather not recollect, but which caused my father great emotion. Less than two years later, he left us.

2

Some Philosophical Reflections on de Finetti's Thought

PATRICK SUPPES

My reflections are divided into four parts. The first deals with de Finetti's pragmatism, the second with his rejection of determinism and indeterminism, the third with the problem of axiomatizing qualitative probability, and the fourth on coherence and consistency.

1   De Finetti's pragmatism

Throughout his life, as far as I have a good sense of that, de Finetti held to a strong form of pragmatism. Here is an important quotation from him in his later years from the translation of his *Theory of Probability* (de Finetti 1975: vol. 2, 201):

> In the philosophical arena, the problem of induction, its meaning, use and justification, has given rise to endless controversy, which, in the absence of an appropriate probabilistic framework, has inevitably been fruitless, leaving the major issues unresolved. It seems to me that the question was correctly formulated by Hume (if I interpret him correctly – others may disagree) and the pragmatists (of whom I particularly admire the work of Giovanni Vailati.)

And here is de Finetti's footnote on Vailati (de Finetti 1975: vol. 2, 201):

> Cf. G. Vailati, *Scritti* (Edited by Seeber), Florence (1911). Giovanni Vailati, a mathematician of the Peano school, was an original, profound and committed supporter of pragmatism in Italy

Bruno de Finetti, Radical Probabilist
Maria Carla Galavotti (ed.)
Copyright © 2009

(which had several features – which I, in fact, approve of – distinguishing it from the American version of Peirce, James, etc.). ...

Let me say something about Giovanni Vailati. He admired Peirce and James, but the form of pragmatism he developed in Italy had several features that distinguished it from the American version. Here is a quotation from Vailati in 1909:

> The term "pragmatism," according to its original creator Ch. S. Peirce, appeared for the first time in 1871, in a series of debates between the members of the Metaphysical Club in Cambridge, Mass. Peirce found that this was the proper word to indicate the method followed by Berkeley in his investigations of the concepts of "substance," "matter," "reality," etc. – even if such a method was not explicitly formulated by the author. ... Peirce thought that this procedure used by Berkeley was an instance of a more general methodological process, which could be described like this: *the only way to determine and clarify the [meaning] of an assertion is to indicate which particular experiences, according to such assertion, are going to take place, or would take place given specific circumstances.* ... "Pragmatism" can be conceived to have a "utilitarian" character only to the extent that it makes it possible to get rid of a certain number of "useless" issues... For instance, when we have two assertions and we are not able to identify the particular experiences that should occur in order to make one of the assertions true and not the other, it is not proper to inquire which of the two is true. In a case like this the two assertions, according to Peirce, have to be considered simply as two different ways to say the same thing. ... Peirce's methodological rule appears to be, for what has been said so far, an indication of the importance of examining our assertions to identify a part that implies some predictions, because that is the part that can be confirmed or refuted by further experiences. ... To entertain a certain belief instead of another means this, for the pragmatist: to have a certain kind of expectations, different from the expectations he would have, having he had a different belief. ... In fact, besides our beliefs regarding the future, we have at least as many beliefs that, apparently, are only facts of the present or the past. Nevertheless, if we look closer at such beliefs we can see that a reference to the future is always an essential part of their meaning. A typical example of this, examined by Berkeley, is represented by judgments on the existence of material objects. In his

Theory of Vision – that is, after all, in every respects a theory of "prediction" [previsione]. The common opinion is that the size, position and distance of objects are perceived in the same way as we perceive their color. Instead our visual perceptions are not able to provide this kind of information immediately. Distances, shapes, and dimensions of objects are not "seen" by us, but rather "foreseen", or inferred from the signs provided by actual visual perceptions. ... (Vailati 1909, trans. by Claudia Arrighi)

This long passage is especially marked by its orientation of pragmatism towards predicting the future. One cannot help but think that de Finetti's own theory of prevision was much influenced by what Vailati has to say, especially in the last paragraph focused on Berkeley. It is not possible to sketch all the ways in which de Finetti's thought was influenced by Vailati, but it is quite clear that there is a natural sympathy of ideas here evident to any careful reader. It is unfortunate that the work of the Italian pragmatists is not more available in English.

One way of offering more detail is to examine de Finetti's allegiance to operationalism, exemplified earlier in the work of the American physicist P.W. Bridgman, explicitly referenced by de Finetti in several places, but especially in the final chapter of his 1937 Paris lectures. He refers there to Bridgman's influential work *The Logic of Modern Physics*, published in 1927.

Here is de Finetti's own very clear statement about operationalism:

With each of our assertions, a question invariably surges into our mind: has this assertion really any meaning? To give only one example, we know that the notion of simultaneity seemed, not very long ago, perfectly clear and sure, to the point that it had been thought possible to consider time as a notion given *a priori*. Why do we no longer believe this today? Because we have been taught the necessity of conceiving of every notion from a point of view which can be called "operational". Every notion is only a word without meaning so long as it is not known how to verify practically any statement at all where this notion comes up; in the example given above, this practical verification is furnished us by Einstein's procedure employing light signals. An analogous evolution took place some time ago in the mathematical sciences: once, for example, the problem of knowing if $1 - 1 + 1 - 1 + \ldots = \frac{1}{2}$ or not was considered in a nebulous, mysterious, metaphysical way; it sufficed to define what was to be understood by "limit" (for example ordinary limit, limit in the sense of Cesaro) and all the obscurities vanished. (de Finetti 1937/1964: 148)

Later, de Finetti goes on to say, "We only apply the notion of probability in order to make likely predictions" (p. 150). This additional focus almost exclusively on prediction takes us immediately back to Vailati's form of pragmatism. I think it is fair to say that of all the major figures in the foundations of probability, it is de Finetti who is most deeply committed to pragmatism and operationalism as the general philosophical foundations of his thought.

2   Rejection of determinism and indeterminism

Again, let me begin with a quotation from de Finetti, an important one, also from vol. 2 of *The Theory of Probability*:

> Before turning to another topic, it would perhaps be appropriate to clarify certain views on the theme of determinism, given the connection with discussions pertaining to the present theory, and given that we have commented upon it (even though in order to decide that it was not relevant). In my opinion, the attachment to determinism as an *exigency of thought* is now incomprehensible. Both classical statistical mechanics (or Mendelian hereditary) and quantum physics provide explanations – in the form of coherent theories, accepted by many people – of apparently deterministic phenomena. The mere existence of such explanations should be sufficient to give the lie for evermore to the dogmatism of this point of view. (de Finetti 1975: vol. 2, 324)

Almost everyone who has read very much of de Finetti at all knows very well his strong rejection of determinism, well stated in the passage that I have quoted, but to be found in many other places in his writing. I'll have more to say about this later. Here is another quote stating his animosity to determinism in another way, bringing out, in this case, his sympathy with the views of von Neumann:

> The foundations of physics are those we have today (perhaps for many decades, perhaps centuries), and I think it unlikely that they can be interpreted (or adapted) in deterministic versions, like those that are apparently yearned for by people who invoke the possible existence of 'hidden parameters', or similar devices. I hold this view not only because von Neumann's arguments against such an

idea seem to me convincing (vN, pp. 313-328), but also because I can see no reason to yearn for such a thing, or to value it – apart from an anachronistic and nostalgic prejudice in favour of the scientific fashion of the nineteenth century. If anything, I find it, on the contrary, distasteful; it leaves me somewhat bewildered to have to admit that the evolution of the system (i.e., of its functions ψ) is deterministic in character (instead of, for example, being a random process) so that indeterminism merely creeps in because of the observation, rather than completely dominating the scene. (de Finetti 1975: vol. 2, 325)

Maria Carla Galavotti has brought to my attention the important fact that de Finetti rejected any absolute or metaphysical concept of indeterminism, as well as that of determinism. Here is a revealing quotation, published late in de Finetti's career:

The alternative [between determinism and indeterminism] is undecidable and ... illusory. These are metaphysical diatribes over 'things in themselves'; science is concerned with what 'appears to us', and it is not strange that, in order to study these phenomena it may in some cases seem more useful to imagine them from this or that standpoint, by means of deterministic theories... or indeterministic ones. (de Finetti 1976: 299-300)

For a more extended discussion of this topic, see Galavotti (1989). Rather than pursue this general topic, I prefer to sketch a view about determinism and indeterminism that is not de Finetti's, but is one with which I think he would be in much agreement if it were laid out in detail, with the appropriate theorems that I refer to. A good starting point is the entropy of familiar Bernoulli or Markov stochastic processes. I do not write down the equation, but note that the notion of entropy here is the rate of entropy change, not the entropy at a cross-sectional moment of a stationary process. The beginning of this story is the theorem proved by Kolmogorov and Sinai in 1958:

THEOREM 1. (Kolmogorov 1958, Kolmogorov 1959 and Sinai 1959). *If two Bernoulli or Markov processes are isomorphic then their entropies are the same.*

Until they proved this theorem the seemingly simple problem was open, whether or not the Bernoulli process with probability 1/2 heads and 1/2 tails, and the Bernoulli process of a three-sided coin, with probability 1/3 for each side, are isomorphic. The theorem proves that they are not, because their entropies are not the same, and therefore by contraposition, they are not isomorphic. The more difficult and surprising converse theorem was proved by Ornstein in 1970 for the case of Bernoulli processes:

THEOREM 2. (Ornstein 1970). *If two Bernoulli processes have the same entropy they are isomorphic.*

In a rather short time, and in a rather straightforward way, Theorem 2 was generalized to:

THEOREM 3. *Any two irreducible, stationary, finite-state discrete Markov processes are isomorphic if and only if they have the same periodicity and the same entropy.*

And, this theorem has as a natural corollary the following: an irreducible, stationary, finite-state discrete Markov process is isomorphic to a Bernoulli process of the same entropy if and only if it is aperiodic. The surprising and important development, particularly with the proof of Ornstein's theorem in 1970, is that entropy is a complete invariant for the measure-theoretic isomorphism of ergodic Bernoulli or Markov processes. The word "complete" here means that to know if two such processes are isomorphic, we need only know if a single number is the same for both of them, namely the entropy rate.

Now I want to apply these ideas to a new view of determinism and indeterminism. A good place to begin is the following quote from Peirce in 1892:

> Try to verify any law of nature, and you will find that the more precise your observations, the more certain they will be to show irregular departures from the law. We are accustomed to ascribe these, and I do not say wrongly, to errors of observation; yet we cannot usually account for such errors in any antecedently probable way. Trace their causes back far enough, and you will be forced to admit they are always due to arbitrary determination, or chance. (Peirce 1892/1955: 331).

Peirce is making, in some ways, an obvious but still very important point. The verification of deterministic laws can scarcely ever be said to be complete, because there will be errors of measurement if continuous quantities are involved. And, even in discrete processes involving large numbers, errors are almost always found in the collection and analysis of data. It is a way of saying that the notion of a strictly deterministic law is an idealized characterization of scientific laws.

I now turn to the beautiful example of Sinai billiards (named after the Russian mathematician Ya. G. Sinai for his important results, 1970). They are so named, because, unlike the idealized deterministic periodic motion of an idealized particle of classical mechanics, chaos is introduced by placing a convex object in the middle of the billiard table. The particle is reflected off this convex object according to the same physical laws as for the sides of the table. So, particles as idealized billiard balls satisfy the following rules: we have a rectangular box with reflecting sides, the classical law of reflection holds: the angle of reflection equals the angle of incidence, and there is no dissipation of energy. For Sinai billiards the convex obstacle is also an ideal reflector. For this physical situation we have the following idealized theorem: the motion of a Sinai billiard ball is ergodic, and as a corollary of that such motion is strongly chaotic.

It is widespread folklore in discussions of chaos by physicists that most important physical examples of chaos are deterministic. On the other hand, there is a variety of evidence, especially mathematical arguments, that associated with chaos, particularly in the strongest chaotic examples, are phenomena that can only be regarded as genuinely random or stochastic in nature. It would be easy to argue that one has got to choose either the deterministic or stochastic view of phenomena, and at least for a given set of cases, it is not possible to move back and forth in a coherent fashion. It is this view, also perhaps part of the folklore, that we want to argue very much against in the present discussion. We will be depending on general ideas from ergodic theory and in particular on the strong kind of isomorphism theorems proved by Donald Ornstein and his colleagues. Before we turn to the details, there are one or two other points we want to discuss in a very intuitive fashion. For example, if we use a billiard model of a mechanical particle, and we consider the deterministic model in the case of an ergodic motion, that is, one, for example, where there is a convex obstacle in the middle of the billiard table, then there is an empirically indistinguishable stochastic model. The response to this isomorphism might be, "Well yes, but for the case of ergodic motion where the convex object is present we should choose either the simple Newtonian model". Because this Newtonian model works so well in the nonergodic periodic case when there is no convex object, it is natural to say that it is not a real choice between the deterministic or stochastic models. Because of its generalizability the choice seems obviously to be the deterministic model.

But this argument can run too far and into trouble when we turn to a wider set of cases. On the same line we would be pushed to argue that the only kind of complete physical model for quantum mechanics must be a deterministic one, for example the kind advocated by Bohm, but the evidence, once we turn to quantum mechanical phenomena, seems far from persuasive for selecting as the unique intuitively correct model the deterministic one. Here there is much to be said for choosing the stochastic model, which is much closer in spirit to the standard interpretation of classical quantum mechanics. Our point, without going into details, is that whether we intuitively believe the model should be deterministic or stochastic will vary with the particular physical phenomena we are considering. What is fundamental is that independent of this variation of choice of examples or experiments is that when we do have chaotic phenomena, especially when we have ergodic phenomena, then we are in a position to choose either a deterministic or stochastic model. When such a choice between different models has occurred previously in physics – and it has occurred repeatedly in a variety of examples, such as free choice of a frame of reference in Galilean relativity, or choice between the Heisenberg or Schroedinger representation of quantum mechanics –, the natural move is toward a more abstract concept of invariance. What is especially interesting about the empirical indistinguishability and the resulting abstract invariance in the present billiard case, is that at the mathematical level the different kinds of models are inconsistent, that is, the assumption of both the deterministic and the stochastic model leads to a contradiction when fully spelled out. On the other hand, it leads to no contradiction at the level of observations, as we shall see in an important class of ergodic cases. (The remarks in the preceding paragraph and this one are close to ones made several years ago in a joint article with Acacio de Barros (Suppes and de Barros 1996).)

We can go further in terms of Sinai billiards with the concept of measure-theoretic isomorphism. To keep things in the context of finite-state discrete processes, we can form a finite partition of the free surface on the billiard table. This constitutes a finite partition of the space of possible trajectories for the billiard and we correspondingly make time discrete in terms of movement from one element of the partition to another. With these constructive approximations, the following theorem has been proved:

THEOREM 4. (Gallavotti and Ornstein 1974). *With the discrete approximation by a finite partition of the continuous flow just described above, the discrete deterministic model of the billiard is isomorphic in the measure-theoretic sense to a finite-state discrete Bernoulli process model of the motion of the billiard.*

It should be noted that instead of this theorem, we could have stated a theorem for continuous time and such results are to be found in the paper by Gallavotti and Ornstein. What the Gallavotti and Ornstein theorem shows is that the discrete mechanics of billiard balls is in the measure-theoretic sense isomorphic to a discrete Bernoulli analysis of the same phenomena. However, it is to be emphasized that in order to claim that intuitively the two kinds of analysis are observationally indistinguishable we need a stricter concept of isomorphism.

To show why this is so, we do not have to consider something as complicated as the billiard example but only compare a first-order Markov process and a Bernoulli process that have the same entropy rate and therefore are isomorphic in the measure-theoretic sense. It is easy to show by very direct statistical tests whether a given sample path of any length, which is meant to approximate an infinite sequence, comes from a Bernoulli process or a first-order Markov process with strong transition dependencies. There is, for example, a simple chi-square test for distinguishing between the two. It is a test for first-order versus zero-order dependency. The analysis is statistical and, of course, cannot be inferred from a single observation, but the data are usually decisive even for finite sample paths that consist of no more than 100 or 200 trials.

A natural stricter concept is that of α-congruence due to Ornstein and Weiss (1991). My explanation is intuitive in terms of trajectories of Sinai billiard balls. Let D_t be a classical mechanics deterministic model with α the bound away from zero of errors of measurement, and let S_t be a stochastic model such that the predictions of the trajectory of a Sinai billiard such that D_t and S_t satisfy

(i) Both models predict correctly the trajectory within the error bound α.

(ii) The theoretical predicted trajectories τ of models D_t and S_t are such that for any two points x and y at time t, the distance between $\delta_t(x,y)$ (deterministic model) and $\sigma_t(x,y)$ (stochastic model) is less than α, except for a set of exceptional points whose probability of occurring is less than α.

So this concept of congruence is a probabilistic one, as a generalization of the familiar Euclidean distance between points in a plane. Using α-congruence the following remarkable theorem can be proved.

THEOREM 5. (Ornstein and Weiss 1991). *There are physical processes which can equally well be analyzed as deterministic systems of classical mechanics or as indeterministic Markov processes, no matter how many observations are made, if observations have an accuracy bounded away from zero.*

I think de Finetti would have been pleased that for important chaotic examples of physical systems there is no observable distinction between deterministic and indeterministic theories of the physical systems. There being no observable distinction resonates nicely with his pragmatism and operationalism.

Here is a very qualitative description of the kind of Markov process cleverly constructed by Ornstein and Weiss to satisfy their Theorem.

1. Pick $\alpha \succ 0$.
2. Define a (biased) coin-tossing stochastic process.
3. Finitely partition the table.
4. The ball will always be in one element of the finite partition.
5. It stays in each element p of the partition for time $t(p)$.
6. The ball then jumps to one of a pair of points according to a toss of the coin.
7. This pair of points depend on p.

Here are some final remarks on Ornstein and Weiss' results that are meant to be in de Finetti's spirit.

1. Such explicit theorems for concrete systems are difficult, but seem likely to be true for a wide variety of chaotic phenomena.

2. Such results are important in making the case that any thesis of universal determinism or indeterminism is transcendental i.e., beyond experience.

3. Transcendental, not false.

3   Axiomatizing qualitative probability

One of the things for which de Finetti is justly famous is his emphasis on the qualitative nature of probability and the possibility of expressing many of our ordinary ideas about probability in qualitative as well as subjective terms (de Finetti 1937/1964). These ideas are expressed well in a long passage from the famous 1937 Paris lectures, which I quote:

Let us consider the notion of probability as it is conceived by all of us in everyday life. Let us consider a well-defined event and suppose that we do not know in advance whether it will occur or not; the doubt about its occurrence to which we are subjects lends itself to comparison, and, consequently, to gradation. If we acknowledge only, first, that one uncertain event can only appear to us (a) equally probable, (b) more probable, or (c) less probable than another; second, that an uncertain event always seems to us more probable than an impossible event and less probable than a necessary event; and finally, third, that when we judge an event E' more probable than an event E, which is itself judged more probable than an event E'', the event E' can only appear more probable than E'' (transitive property), it will suffice to add to these three evidently trivial axioms a fourth, itself of a purely qualitative nature, in order to construct rigorously the whole theory of probability. This fourth axiom tells us that inequalities are preserved in logical sums: if E is incompatible with E_1 and with E_2, then $E_1 \vee E$ will be more or less probable than $E_2 \vee E$, or they will be equally probable, according to whether E_1 is more or less probable than E_2, or they are equally probable. More generally, it may be deduced from this that two inequalities, such as

E_1 is more probable than E_2,

E_1' is more probable than E_2',

can be added to give

$$E_1 \vee E_1' \text{ is more probable than } E_2 \vee E_2',$$

provided that the events added are incompatible with each other (E_1 with E_1', E_2 with E_2'). It can then be shown that when we have events for which we know a subdivision into possible cases that we judge to be equally probable, the comparison between their probabilities can be reduced to the purely arithmetic comparison of the ratio between the number of favorable cases and the number of possible cases (not because the judgment then has an objective value, but because everything substantial and thus subjective is already included in the judgment that the cases constituting the division are equally probable). This ratio can then be chosen as the appropriate index to measure a probability, and applied in general, even in cases other than those in which one can effectively employ the criterion that governs us there. In these other cases one can evaluate this index by comparison: it will be in fact a number, uniquely determined, such that to numbers greater or less than that number will correspond events respectively more probable or less probable than the event considered. Thus, while starting out from a purely qualitative system of axioms, one arrives at a quantitative measure of probability, and then at the theorem of total probability which permits the construction of the whole calculus of probabilities. (de Finetti 1937/1964: 100-101)

In Definition 1, I express these axioms numbered as by de Finetti, with only slight changes. Axiom D0 just defines the formal structures used, consisting of a nonempty set Ω, an algebra of sets representing events closed under union and complementation, and the qualitative relation $\pm$ expressing (weakly) more probable than:

DEFINITION 1. A *structure* $\Omega = (\Omega, \Im, \pm)$ *is a qualitative probability structure if and only if the following axioms are satisfied for all A, B, and C in* $\Im$:

D0. $\Im$ *is an algebra of sets on* Ω;

D1. $A \pm B$ *or* $B \pm A$;

D2. *If* $A \neq \emptyset$, *then* $A \succ \emptyset$ *and* $\Omega \pm A$.

D3. *If* $A \pm B$ *and* $B \pm C$, *then* $A \pm C$;

D4. *If* $A \cap C = \emptyset$ *and* $B \cap C = \emptyset$, *then* $A \pm B$ *if and only if* $A \cup C \pm B \cup C$.

De Finetti's subdivision remark after formulating the fourth axiom, the one expressing additivity, is one way to get immediately a quantitative representation in the finite case. For this purpose we define equivalence ≈ of events in the standard way: $A ≈ B$ if and only if $A \pm B$ and $B \pm A$. I will call this Axiom 5, the subdivision axiom, which has a deceptively simple formulation, but it has as a consequence that if the set of possible outcomes is finite, then the possible outcomes all have the same probability.

DEFINITION 2. *Let $Ω$ be a finite set, and let* $(Ω, ℑ, \pm)$ *be a qualitative probability structure. This structure is* uniform *when Axiom 5 is also satisfied*:

AXIOM 5. *If $A \pm B$ then there is a C in $ℑ$ such that $A ≈ B ∪ C$.*

We then have as a straightforward elementary theorem the existence of a unique strictly agreeing probability measure P for such structures when they are finite.

THEOREM 6. *Let $(Ω, ℑ, \pm)$ be a uniform finite qualitative probability structure. Then this structure has a strictly agreeing unique probability measure P, i.e., for events A and B*
$$P(A) ≥ P(B) \text{ if and only if } A \pm B.$$

I stated this theorem and gave the elementary proof in Suppes (1969: 6) before I read de Finetti in any detail. Kraft, Pratt, and Seidenberg (1959) showed that finite qualitative probability structures, in the sense of Definition 1, do not in general have a strictly agreeing probability measure. Their counterexample has five possible outcomes.

There is a further step that de Finetti did not formally take. By having a set of standard events of equal probability, it is possible to approximate the probability of any event even when the space of possible outcomes is infinite.

Here is the definition of the structure and the theorem of approximate measurement of belief I gave in Suppes (1974) with acknowledgment of the earlier work of the godfathers of modern Bayesian thought – Ramsey, de Finetti, and Savage.

From a formal standpoint, the basic structures to which the axioms apply are quadruples $(\Omega, \Im, \mathcal{S}, \pm)$, where as before, Ω is a nonempty set, $\Im$ is an algebra of subsets of Ω, and $\pm$ is the qualitative probability relation, $\mathcal{S}$ is a similar finite algebra of sets, intuitively the events that are used for standard measurements, and I shall refer to the events in $\mathcal{S}$ as *standard events, S, T,* etc.

DEFINITION 3. *A structure* $\Omega = (\Omega, \Im, \mathcal{S}, \pm)$ *is a finite approximate qualitative probability structure if and only if* Ω *is a nonempty set,* $\Im$ *and* $\mathcal{S}$ *are algebras of sets on* Ω, *and the following axioms are satisfied for every A, B and C in* $\Im$ *and every S and T in* $\mathcal{S}$:

  Axiom 1. $(\Omega, \Im, \pm)$ *is a qualitative probability structure in the sense of Definition 1;*
  Axiom 2. $\mathcal{S}$ *is a finite subset of* $\Im$, *and* $(\Omega, \mathcal{S}, \pm)$ *is a uniform qualitative probability structure.*

A minimal element of $\mathcal{S}$ is any event A in $\mathcal{S}$ such that $A \neq \varnothing$, and it is not the case that there is a nonempty B in $\mathcal{S}$ such that B is a proper subset of A. A *minimal open interval* (S, S') of $\mathcal{S}$ is such that $S \prec S'$ and $S' - S$ is equivalent to a minimal element of $\mathcal{S}$. Axiom 5, stated earlier, is the main structural axiom, which holds only for the finite subalgebra and not for the general algebra.

  In stating the representation and uniqueness theorem for structures satisfying Definition 3, in addition to an ordinary probability measure on the standard events, I shall use upper and lower probabilities to express the inexact measurement of arbitrary events. A good discussion of the quantitative properties one expects of such upper and lower probabilities is found in Good (1962). All of his properties are not needed here because he dealt with conditional probabilities. The following properties are fundamental, where $P_*(A)$ is the lower probability of an event A and $P^*(A)$ is the upper probability (for every A and B in $\Im$):

  I. $P_*(A) \geq 0$.
  II. $P_*(\Omega) = P^*(\Omega) = 1$.
  III. If $A \cap B = \varnothing$ then

$$P_*(A) + P_*(B) \leq P_*(A \cup B) \leq P_*(A) + P^*(B) \leq P^*(A \cup B) \leq P^*(A) + P^*(B)$$

Condition (I) corresponds to Good's Axiom D2 and (III) to his Axiom D3.

For standard events $P(S) = P_*(S) = P^*(S)$. For an arbitrary event A not equivalent in qualitative probability to a standard event, I think of its "true" probability as lying in the open interval $(P_*(A), P^*(A))$.

In the fourth part of Theorem 7, I define a certain relation and state it is a semiorder with an implication from the semiorder relation holding to an inequality for upper and lower probabilities. Semiorders have been fairly widely discussed in the literature as a generalization of simple orders, first introduced by Duncan Luce. I use here the axioms given by Scott and Suppes (1958). A structure $(A, *\!\succ)$ where A is a nonempty set and $*\!\succ$ is a binary relation on A is a *semiorder* if and only if for all $a, b, c, d \in A$:

Axiom 1. Not $a * \succ a$;

Axiom 2. If $a *\!\succ b$ and $c *\!\succ d$ then either $a *\!\succ d$ or $c *\!\succ b$;

Axiom 3. If $a *\!\succ b$ and $b *\!\succ c$ then either $a *\!\succ d$ or $d *\!\succ c$.

THEOREM 7. *Let $\Omega = (\Omega, \Im, \mathcal{S}, \pm)$ be a finite approximate qualitative probability structure. Then*

(i) there exists a probability measure P on $\mathcal{S}$ such that for any two standard events S and T

$$S \pm T \text{ if and only if } P(S) \geq P(T),$$

(ii) the measure P is unique and assigns the same positive probability to each minimal event of $\mathcal{S}$,

(iii) if we define P_* and P^* as follows:

(a) for any event A in $\Im$ equivalent to some standard event S,

$$P_*(A) = P^*(A) = P(S),$$

(b) for any A in $\Im$ not equivalent to some standard event S, but lying in the minimal open interval (S, S') for standard events S and S'

$$P_*(A) = P(S) \text{ and } P^*(A) = P(S'),$$

then P_* and P^* satisfy conditions (I)--(III) for upper and lower probabilities on $\Im$, and

(c) if n is the number of minimal elements in $\mathcal{S}$ then for every A in $\Im$

$$P^*(A) - P_*(A) \leq 1/n,$$

(iv) if we define for A and B in $\mathfrak{I}$

  $A *\succ B$ if and only if $\exists S$ in $\mathfrak{S}$ such that $A \succ S \succ B$,

then $*\succ$ is a semiorder on $\mathfrak{I}$, if $A *\succ B$ then $P_*(A) \geq P^*(B)$, and if
$P_*(A) \geq P^*(B)$ then $A \pm B$.

This theorem expresses a simple constructive result about approximate measurement of subjective probability. It is, I believe, very much in the spirit of informal remarks that occur in de Finetti's writings about such approximations.

Finally, I consider an extension of Definition 1 to give necessary and sufficient conditions for the existence of a unique strictly agreeing probability measure. In brief, by enlarging the structures to include elementary random variables, Definition 1 can be extended to give necessary and sufficient conditions for all sets Ω, finite or infinite.

If A is a set, $\mathbf{A}^i$ is its indicator function, which is a random variable. Thus, if A is an event

$$\mathbf{A}^i(\omega) = \begin{cases} 1 & \text{if } \omega \in A, \\ 0 & \text{otherwise.} \end{cases}$$

We need to go slightly beyond the indicator functions. The move is from an algebra of events to the algebra $\mathfrak{I}^*$ of extended indicator functions relative to $\mathfrak{I}$. The algebra $\mathfrak{I}^*$ is just the smallest semigroup (under function addition) containing the indicator functions of all events in $\mathfrak{I}$. In other words, $\mathfrak{I}^*$ is the intersection of all sets with the property that if A is in $\mathfrak{I}$ then $\mathbf{A}^i$ is in $\mathfrak{I}^*$ and if $\mathbf{A}^*$ and $\mathbf{B}^*$ are in $\mathfrak{I}^*$, then $\mathbf{A}^* + \mathbf{B}^*$ is in $\mathfrak{I}^*$.

Then, to have $\mathbf{A}^* \pm \mathbf{B}^*$ is to have, intuitively, the expected value of $\mathbf{A}^*$ equal to or greater than the expected value of $\mathbf{B}^*$. The qualitative comparison is now not one about the probable occurrences of events, but about the qualitative expected value of certain restricted random variables. What the representation theorem below shows is that very simple necessary and sufficient conditions on the qualitative comparison of extended indicator functions guarantee existence of a strictly agreeing, finitely additive measure P, whether the set Ω of possible outcomes is finite or infinite, i.e., $P(A) \pm P(B)$ if and only if $A \pm B$.

DEFINITION 4 (Suppes and Zanotti 1976). *Let Ω be a nonempty set, let $\Im$ be an algebra of sets on Ω, and let $\pm$ be a binary relation on $\Im^*$, the algebra of extended indicator functions relative to $\Im$. Then the qualitative algebra $(\Omega, \Im, \pm)$ is qualitatively satisfactory if and only if the following axioms are satisfied for every $\mathbf{A}^*, \mathbf{B}^*, and \mathbf{C}^*$ in $\Im^*$:*

   Axiom 1. *The relation $\pm$ is a weak ordering of $\Im^*$;*
   Axiom 2. $\Omega^i \succ \varnothing^i$;
   Axiom 3. $\mathbf{A}^* \pm \varnothing^i$;
   Axiom 4. $\mathbf{A}^* \pm \mathbf{B}^*$ *if and only if* $\mathbf{A}^* + \mathbf{C}^* \pm \mathbf{B}^* + \mathbf{C}^*$;
   Axiom 5. *If* $\mathbf{A}^* \succ \mathbf{B}^*$ *then for every* $\mathbf{C}^*$ *and* $\mathbf{D}^*$ *in* $\Im^*$ *there is a positive integer n such that* $n\mathbf{A}^* + \mathbf{C}^* \pm n\mathbf{B}^* + \mathbf{D}^*$.

These axioms in terms of qualitative expectation fit in with de Finetti's framework of analysis. Only Axiom 5 needs explanation. It is one standard form of an Archimedean axiom.

THEOREM 8. *Let Ω be a nonempty set, let $\Im$ be an algebra of sets on Ω, and let $\pm$ be a binary relation on $\Im$. Then a necessary and sufficient condition that there exists a strictly agreeing probability measure on $\Im$ is that there be an extension of $\pm$ from $\Im$ to $\Im^*$ such that the qualitative algebra of extended indicator functions $(\Omega, \Im^*, \pm)$ is qualitatively satisfactory.*

Moreover, if $(\Omega, \Im^*, \pm)$ is qualitatively satisfactory, then there is a unique strictly agreeing expectation function on $\Im^*$ and this expectation function generates a unique strictly agreeing probability measure on $\Im$.

4   Coherence and consistency

I summarize my main points, which, I realize, not everyone will agree with.

1. Coherence is an important concept for de Finetti and almost all subjectivists.

2. Prior to experimentation, coherence replaces truth as the central philosophical concept.

3. But coherence is not a strong enough constraint on experimental scientists or engineers prior to actual experimentation. We must believe their priors are based on serious past experience. This is a point not well enough recognized by some Bayesians.

4. It is reasonable to hold that even for the hypothesis being tested, "Not all priors are equal."

5. My focus is on coherence itself and its formal complexity as a problem.

6. In most advanced work in statistics, the problem of coherence is assumed, not investigated. It amounts to assuming the family of random variables for the problem at hand has a joint distribution, even when the prior information is far from being decisive on this point. As an example, the nonexistence of joint distributions for pairwise correlations is a familiar aspect of quantum-entanglement experiments.

7. As Fréchet pointed out, it is easy for persons to have irrational, i.e., incoherent probabilities for complex situations.

8. Here is a simple artificial example. We are given three random variables X, Y, and Z with possible values ± 1 and expectations

$$E(X) = E(Y) = E(Z) = 0,$$
and

$$E(XY) = 0.6$$
$$E(YZ) = 0.7$$

Given a meteorological story or something similar, respondents are asked their prior for $E(XZ) = ?$

Bob says, I estimate $E(XZ) = 0.25$.

Question: Is Bob coherent? Answer: No.

We have the following inequalities for the three correlations of X, Y, and Z to have a joint distribution (Suppes and Zanotti 1981):

$$-1 \leq E(XY) + E(YZ) + E(XZ) \leq 1 + 2Min(E(XY), E(YZ), E(XZ)).$$

The values assigned to the correlations do not satisfy these inequalities, so Bob is incoherent, even though he is unlikely to know it. An objection to this example is that I am assigning priors to the correlations, but this is really no different than asking for the three pairwise distributions, and usually the triple moment $E(XYZ)$ would be ignored. The essential point is that there is little if any discussion in the literature of complicated priors and their problem of coherence. Betting quotients are not constructed in fact for anything very complicated. I am just expressing here my skepticism that for complicated stochastic processes and related entities, proof of coherence can be taken as a serious requirement, any more than proof of consistency is a prerequisite to do classical mathematical analysis.

9. Consistency has been an ideal but unattainable goal of pure mathematics since the revolutionary results of Gödel in the 1930s. Based on this experience, coherence seems hard to guarantee in advanced work in probability and statistics, a point that does not seem to be fully appreciated among Bayesian statisticians. I am just sorry I did not ever discuss this problem with de Finetti, on the several occasions when we had long philosophical conversations. But I do want to make clear I think he would have had a lot to say, and would have had no difficulty in putting his own touch on how to think about such problems. Here is a quotation from 1961, the publication of the proceedings of a colloquium entitled *La Décision* in 1960 in Paris, the occasion of my first meeting with de Finetti:

> Sans doute, dans les évaluations pratiques complexes tout homme réel est incapable d'échapper à des contradictions. Malheureusement, la qualification de "comportement rationnel" employée parfois, assez improprement, pour le comportement conforme à la théorie, a fait souvent soupçonner qu'on prétendait que tous les hommes (les fous exceptés) seraient infailliblement et automatiquement conduits par leur propre psychologie à s'y tenir; ou que c'est la psychologie que l'on a imaginée pour un type d'homme hautement idéalisé. Ni l'un ni l'autre, comme on vient de voir.

> Un fait différent est par contre d'admettre explicitement comme un des effets de l'idéalisation propre à tout schème: mes décisions réelles dépendent aussi de facteurs accessoires qu'il faut considérer à côté du schème principal. (de Finetti 1961: 164)

We had at that 1960 meeting a lively discussion of the special role of the axiom of choice in the foundations of mathematics, but I can no longer remember any of the particular remarks either one of us made.

The details are gone from our several meetings over two decades, but even now, more than twenty years after de Finetti's death, I remember vividly one lasting impression. Of the many mathematicians and statisticians I have known over my long life, Bruno de Finetti was the most deeply philosophical.

References

Bridgman, P.W. (1927). *The Logic of Modern Physics*. New York: Macmillan.

de Finetti, B. (1931). "Sul Significato Soggettivo della Probabilità". *Fundamenta Mathematicae* XVII, pp. 298-329. English edition: "On the Subjective Meaning of Probability". English translation in B. de Finetti, *Probabilità e Induzione (Induction and Probability)*, ed. by P. Monari and D. Cocchi. Bologna: CLUEB, 1992, pp. 291-321.

de Finetti, B. (1937/1964). "La prévision: ses lois logiques, ses sources subjectives". *Annales de l'Institut Henri Poincaré* 7, pp. 1-68. Translated in *Studies in Subjective Probability*, ed. by H.E. Kyburg, Jr. and H.E. Smokler. New York: Wiley, 1964, pp. 93-158.

de Finetti, B. (1961). "Dans quel sens la théorie de la decision est-elle et doit-elle être 'normative'?". In *La Décision*. Paris: Éditions Du Centre National De La Recherche Scientifique, pp. 159-169.

de Finetti, B. (1975). *Theory of Probability*, vol. 2. Translated by A. Machi and A. Smith. New York: Wiley.

de Finetti, B. (1976). "Probability: Beware of Falsifications!". *Scientia* 70, pp. 299-300.

Galavotti, M.C. (1989). "Anti-realism in the Philosophy of Probability: Bruno de Finetti's Subjectivism". *Erkenntnis* 31, pp. 239-261.

Gallavotti, G. and Ornstein, D.S. (1974). "Billiards and Bernoulli Schemes". *Comm. Math. Phys.* 38, pp. 83-101.

Good, I.J. (1962). "Subjective Probability as the Measure of a Non-measurable Set". In *Logic, Methodology, and Philosophy of Science: Proceedings of the 1960 International Congress*, ed. by E. Nagel, P. Suppes and A. Tarski. Stanford: Stanford University Press, pp. 319-329.

Kolmogorov, A.N. (1958). "A New Metric Invariant of Transient Dynamical Systems and Automorphisms in Lebesgue Spaces". *Dokl. Akad. Nauk. SSSR* 119, pp. 861-864. (Russian) MR 21 #2035a.

Kolmogorov, A.N. (1959). "Entropy per Unit Time as a Metric Invariant of Automorphism. *Dokl. Akad. Nauk SSSR* 124, pp. 754-755. (Russian) MR 21 #2035b.

Kraft, C.H., Pratt, J.W. and Seidenberg, A. (1959). "Intuitive Probability on Finite Sets". *Annals of Math. Statistics* 30, pp. 408-419.

Ornstein, D.S. (1970). "Bernoulli Shifts with the Same Entropy are Isomorphic". *Advances in Mathematics* 4, pp. 337-352.

Ornstein, D.S. and Weiss, B. (1991). "Statistical Properties of Chaotic Systems". *Bull. Am. Math. Soc. (New Series)* 24, pp. 11-116.

Peirce, C. (1892). "The Doctrine of Necessity Examined". In *Philosophical Writings of Peirce*, ed. by J. Buchler. New York: Dover, 1955, pp. 324-338.

Scott, D. and Suppes, P. (1958). "Foundational Aspects of Theories of Measurement". *Journal of Symbolic Logic* 23, pp. 113-128.

Sinai, Ya. G. (1959). "On the Notion of Entropy of a Dynamical System". *Dokl. Akad. Nauk SSSR* 124, pp. 768-771.

Sinai, Ya. G. (1970). "Dynamical Systems with Elastic Reflections: Ergodic Properties of Displacing Billiards". *Dokl. Akad. Nauk SSSR* 124, pp. 137-189.

Suppes, P. (1969). *Studies in the methodology and foundations of science: Selected papers from 1951-1969*. Dordrecht: Reidel.

Suppes, P. (1974). "The Measurement of Belief". *Journal of the Royal Statistical Society (Series B)* 36, pp. 160-191.

Suppes, P. (2006). "Transitive Indistinguishability and Approximate Measurement with Standard Finite Ratio-scale Representations". *Journal of Mathematical Psychology* 50, pp. 329-336.

Suppes, P. and de Barros, J.A. (1996). "Photons, Billiards and Chaos". In *Law and Prediction in the Light of Chaos Research, Lecture Notes in Physics*, ed. by P. Weingartner and G. Schurz. Berlin: Springer-Verlag, pp. 189-201.

Suppes, P. and Zanotti, M. (1976). "Necessary and Sufficient Conditions for Existence of a Unique Measure Strictly Agreeing with a Qualitative Probability Ordering". *Journal of Philosophical Logic* 5, pp. 431-438.

Suppes, P. and Zanotti, M. (1981). "When are Probabilistic Explanations Possible?". *Synthese* 48, pp. 191-199.

Vailati, G. (1909). "The Origins and Fundamental Idea of Pragmatism". *Rivista di Psicologia Applicata* 1. Reprinted in *Scritti di G. Vailati*. Leipzig: J.A. Barth, 1911, pp. 36-48.

3

Logic and Finite Additivity: Mutual Supporters in Bruno de Finetti's Probability Theory
COLIN HOWSON

1 Introduction

That a great thinker should be celebrated in his lifetime is not particularly unusual, but that Bruno de Finetti, a mathematician, statistician and economist, should have become very celebrated indeed for what most people would regard as a fundamental contribution to epistemology is fairly remarkable. The problems that he addressed, and far as many people are concerned solved, are in the notoriously intractable philosophical area known generically as *the problem of induction*. Everyone who wants to understand the contemporary state of debate on this problem will have at some point to become acquainted with the intellectual tools that de Finetti brought to the discussion, and which have wrought a revolution in the way we think of the problem[1]. These are the concepts of *subjective probability*, *coherence*, *Dutch Book argument*, and *exchangeability*.

2 Coherence

De Finetti's discussion of coherence appears to be a model of mathematical clarity and precision. It is also well-known, but since it is the *mis-en-scène* of this paper I will briefly recapitulate it. De Finetti describes two operational methods for eliciting a subject's probability-valuations, and

---

[1] Very good introductions to de Finetti's philosophy of probability are to be found in Gillies (2000) and Galavotti (2005).

Bruno de Finetti, Radical Probabilist
Maria Carla Galavotti (ed.)
Copyright © 2009

criteria based on each, which he shows to be equivalent in extension, for determining whether the evaluations are coherent. From these criteria, as he shows, the entire calculus of probability can be deduced. The elicitation-methods are: (i) the subject is asked to nominate betting quotients in bets on various sets of propositions $A_1, \ldots, A_n,$[2] subject to the conditions that the subject will bet on or against each A_i, for arbitrary "small" stakes and with the direction of the bet decided by some external authority; (ii) the subject is asked to name his/her probabilities (now called "previsions") $P(A_i)$ of the truth-values of the A_i subject to a penalty equal to $\Sigma[I_{A_i}-P(A_i)]^2$; this quantity is a *quadratic scoring rule*, and is *proper* in the sense that it is in the subject's own interest (in terms of expectation) to nominate his/her genuine evaluations.

Now suppose π is a set of such evaluations according to either (i) or (ii). π is coherent if (in terms of (i)) there is no set of stakes, positive or negative, which assures a loss independently of the truth-values of the Ai (there is no Dutch Book against those evaluations), or (in terms of (ii)), there is no choice of the evaluations $P(A_i)$ which will generate a smaller penalty over all possible truth-values of the A_i (in decision-theoretic language, the choice is admissible: it cannot be dominated by any alternative choice). The definition is legitimate since, as de Finetti showed, the two criteria are extensionally equivalent. There is therefore no loss of generality in focusing on (i), i.e. on the elicitation of the agent's probabilities in terms of the betting quotients at which he/she would be indifferent between simultaneously betting on or against each a finite set of propositions (a suitably modified account will hold for (ii)). These betting quotients determine what the agent is supposed to regard as a fair sum of bets on those propositions (1972: 77).

3   A countable lottery

The betting scenario occupies centre-stage in what is probably de Finetti's best-known paper (1937) which, translated by Kyburg, is included in Kyburg and Smokler's popular anthology on subjective probability (1964, 1980). In that paper de Finetti represents coherence not merely as a prudential constraint but as something much stronger, as nothing less than a condition of *consistency* on one's subjective evaluations, and one

---

[2] A bet with nonzero stake S, positive for a bet on A and negative for a bet against, is a function on possible states of the form $S(I_A(\cdot)-p)$, where p is the betting quotient and $I_A(\cdot)$, the indicator function for A, takes the value 1 on the states making A true and 0 on those making it false.

moreover couched in terms of something like logical consistency. Consider, for example, the following description (as translated by Kyburg) of one whose evaluations are incoherent:

> one clearly should say that the evaluation of the probabilities given by this individual contains an incoherence, *an intrinsic contradiction...* [the probability calculus] then appears as a set of rules to which the subjective evaluation of probability of various events by the same individual ought to conform *if there is not to be a fundamental contradiction among them.* (1937: 63; my emphasis)

In fact "cohérence" (the paper originally appeared in French) and the Italian "coerenza" would normally be translated as "consistency", and the Italian translator of a major paper of de Finetti (1949) renders "coerenza" uniformly as "consistency"[3]. It is not at all obvious, however, why incoherence in the sense of (i) and (ii) should be taken to justify the conclusion that an incoherent distribution should be seen as inconsistent in the way that these remarks suggest.

De Finetti has an answer, however, and it emerges in his discussion of another topic on which his views are well-known if not notorious, that of countable additivity, where his observations appear to be in direct conflict with his account of coherence. The discussion proceeds from the example of a countably infinite lottery, with tickets labelled 1, 2, 3, ... etc., over which there is the uniform probability distribution, $p_n = 0$ for all n, a distribution de Finetti regards as appropriate, in fact mandatory, if the tickets are drawn "at random" (1974: 120). Clearly, the uniform distribution p_i violates countable additivity. Not only that: a countable sum of bets against each ticket being drawn, with stake 1 (1 here just signifies a small unit of some currency), will result in a certain loss of 1 since $p = 0$. In other words, there is a Dutch Book against any owner of this assignment, and a very simple one at that. Only $p = 1$, and hence a strongly asymmetrical distribution, avoids it. Maher sums up the general opinion in concluding that "de Finetti cannot consistently reject countable additivity" (Maher 1993: 200).

But that conclusion is false, as de Finetti himself goes to considerable pains to make clear, in the process also answering the question of why he took incoherent assignments to be actually inconsistent. The key is that

---

[3] It is therefore odd, to say the least, that the translators of de Finetti's book (1974) claim to have followed the translators of that paper and de Finetti's 1937 paper in rendering "coerenza" as "coherence".

the criteria (i) and (ii) above are to be applied to finite assignments only. In the betting scenario of (i), the betting quotients elicited by requiring the agent be indifferent between either side of the bet at the associated odds is usually taken to imply that these bets are regarded by the agent as fair. Since a bet on/against a single proposition is a degenerate finite sum, this amounts to the assumption, call it (*), that fairness for bets with small stakes is closed under finite sums, an assumption de Finetti elsewhere calls the hypothesis of rigidity (1974)[4]. De Finetti's assumption of (*) answers the question why he regarded an agent as genuinely inconsistent in nominating a finite set of fair betting quotients against which a Dutch Book can be constructed: the agent is implicitly declaring fair a gamble (a sum of fair bets) which is vulnerable to a certain loss or gain. That may not be a strictly logical falsehood in the sense of classical formal logic, but given the usual meaning of the vernacular terms it is surely reasonable to regard it as analytically false, as de Finetti maintained (an explicit statement to this effect is in his 1972: 84). And that fairness is closed only under finite sums answers the other question, of why for de Finetti the countable lottery under the uniform distribution is not incoherent/inconsistent despite the fact that it is vulnerable to a Dutch Book. It is not incoherent/inconsistent because without the countable extension of (*) there is no contradiction in simultaneously maintaining that each of an infinite sum of bets is fair and also that the sum itself is unfair.

It remains to tie the discussion, and in particular (*), to finite additivity. De Finetti's objective was to show that violating finite additivity involves the same inconsistency as asserting that a fair sum of bets is uniformly positive (or negative). His famous Dutch Book theorem states that if the domain of a non-negative real-valued function P is a field (or algebra[5]) F, with the probability of the certain event 1, then if finite additivity is violated there is a finite sum X of bets each of which is fair according to P (i.e., P represents the agent's distribution of fair betting quotients over F) such that X>0 (i.e. X(s)>0 for all outcomes s). To prove that vio-

---

[4] Another way of stating the assumption is as that of the *finite additivity of expectations*, or as de Finetti terms them *previsions*, for sufficiently small stakes, which as he points out is a theorem of standard utility theory when the expectations are strictly utility-expectations. Utility is of course a unit of value like money but measured on a scale adjusted for risk-aversion, which approximates the money scale for small sums, and it this approximate correctness for bets with small stakes which justifies (*) (de Finetti 1974: 80-82).

[5] De Finetti follows the usual practice among probabilists of identifying equivalent propositions.

lation of finite additivity implies the inconsistency in question one clearly needs to transform the theorem into the following form: if finite additivity is violated then there is a finite sum X of bets such that X is fair according to P but which is always positive. This is exactly what (*) does (1972: 77-78).

We can now list two important corollaries to the foregoing discussion.

1. Finite additivity is compact, while countable additivity is not. Restricting (*) to finite sums means that there is no inconsistency in having a Dutch Bookable infinite set of probability assignments so long as no finite subset is Dutch Bookable. With finite additivity, therefore, if an assignment is inconsistent then some finite subset is (with countable additivity compactness fails, since if (*) is extended to infinite sums, where they are defined, the countable lottery above is an inconsistent assignment every finite subset of which is consistent). I have deliberately used the word "compact" here because of its deductive associations. I believe, and many of de Finetti's own statements support, the view that he saw the discipline of subjective probability as nothing less than a logic complementary to deductive logic. Thus not only is the title of his famous 1937 paper "La prévision: *ses lois logiques*, ses sources subjectives" (my emphasis) but also, in a paper written shortly before, we read that:

> It is beyond doubt that probability theory can be considered as a multi-valued logic (precisely: with a continuous range of values), and that this point of view is the most suitable to clarify the foundational aspects of the notion and the logic of probability. (de Finetti 1936, parenthesis in the original; quoted in Coletti and Scozzafava (2002: 61). These values were not additional truth-values, but probability-values "superimposed" [sic] on the logic of truth-values.)

Although de Finetti later in his life laid more emphasis on operational criteria with corresponding decision-theoretic justifications, the view that the axioms of probability should be viewed as laws of logic was always close to the surface if not obviously apparent, and when we come to his discussion of the principle of countable additivity we shall find those operational criteria effectively trumped by appeal to what can only be described as fundamentally logical considerations.

2. Using a Dutch Book argument for finite additivity is redundant given (*). It is easy to show that (*) by itself implies finite additivity. Suppose there are countably many disjoint events E_i in the domain F of P

whose union (or disjunction) $\cup E_i$ is also in F. By (*), the sum X of the fair bets $1(I_{Ei}\text{-}P(E_i))$ is fair. But $X = \Sigma I_{Ei} - \Sigma P(E_i) = 1(I_{\cup Ei} - \Sigma P(E_i))$. This is fair just in case $P(\cup E_i) = \Sigma E_i$. A straightforward extension of this argument proves that the countable version of (*) implies countable additivity[6].

3. Dutch Book arguments by themselves prove nothing. As we saw, using Dutch Book arguments to prove incoherence (as a type of genuine inconsistency) presupposes the additivity-of-fairness postulate (*). We have seen that for de Finetti himself there is no inconsistency in maintaining the coherence of the countable fair lottery even though it is vulnerable to a simple Dutch Book, since he repudiates the countable version of (*). As we shall see in Section 4, he insisted on the contrary that in appropriate circumstances the uniform distribution over a countable partition is mandatory. But in that case why stop at infinity (going backwards now!)? Why should the same not go for finite Dutch-Bookable assignments? Why do they not beg the question in the same way?

De Finetti's answer, as we noted, is that with stakes small enough to approximate utilities (*) is a provable consequence of expected utility theory. But this response is problematic, for two reasons. One is that the countable version of (*) is a consequence of a version of Savage's theory to which a suitable axiom of continuity is added (Villegas 1964). De Finetti rejected the countable version of (*) precisely because it entails countable additivity, but this defence of finiteness opens the door to the objection that one might equally argue that (*) itself should be rejected if one finds its consequences objectionable enough. The objection is not merely an academic one, as we shall see. Secondly, even for finite sums the proviso that the stakes be "sufficiently small" is a significant one, since the stakes have to be so small that effects of risk-aversion vanish. But in no practical case will this be true unless they are so small as to invalidate the betting scenario as a reliable way of eliciting degrees of belief, as Ramsey pointed out as a reason for eschewing an approach in terms of money bets altogether (1926: 176). Nor, as de Finetti observed, is it practically possible to invoke pure utility-scaled bets:

---

[6] Hence de Finetti's claim that charging the countable fair lottery with inconsistency *presupposes* countable additivity:

> only if we know that complete additivity holds can we think of extending the notion of combinations of fair bets to combinations of an infinite number of bets, with the corresponding sequence of betting odds. (1972: 91; the crucial word is "fair".)

it would be practically impossible to proceed with transactions, be-cause the real magnitudes in which they have to be expressed ... would have to be adjusted to the continuous and complex varia-tions in a unit of measure [utility] that nobody would be able to ob-serve. (1974: 81)

Even for sums of two bets (*) is still therefore a substantive postulate, and as such one which not only can one consistently reject but also in appropriate circumstances deem false. As far as I am aware the first to point this out in the philosophical literature was Schick; as he observes, the Dutch Book argument for the binary addition principle contains

the unspoken assumption ... that the value I place on [the bets taken] together is the sum of the values I put on them singly. This, however, is not always true – it isn't always true of *me*. (1986: 113)[7]

Quite so. The fact is that (*) is a non-trivial assumption which can be, and in many cases will be, false in any transaction involving actual gambles.

But old ideas die hard. One such is the belief that in the Dutch Book argument de Finetti had produced a powerful and simple tool for validat-ing probability claims. Thus Conditionalisation, both garden-variety and Jeffrey Conditionalisation (aka kinematics), the so-called Reflection Principle, and of course countable additivity are still routinely justified by appeal to Dutch Books arguments, and many regard Ramsey as having put his finger precisely on what is wrong with Dutch Bookable evalua-tions: they violate a fundamental principle of consistency in evaluating the same event/proposition differently:

If anyone's mental condition violated these [probability] laws, his choice would depend on the precise form in which the options were offered him, which would be absurd. He could then have a book made against him by a cunning bettor and would then stand to lose in any event. (1926: 80)

For example Skyrms, quoting the passage, claims that "what is basic [to the Dutch Book argument for the addition principle for probabilities] is the consistency condition that you evaluate a betting arrangement inde-pendently of how it is described" (1984: 21-22). Pointing out that the sum

---

[7] Though as we noted, in the form of postulate (*) it was 'spoken' by de Finetti himself – perhaps rather too quietly, since it is subsequently ignored in the usual accounts of his work.

of two bets on the propositions A and B for a dollar stake with betting quotients equal to your personal probabilities P(A), P(B) is a bet on the disjunction A∨B with the same stake and betting quotient P(A) + P(B), Skyrms concludes (using p, q where I have used A, B)

> if you are to be *consistent*, your personal probability for p *or* q had better be ... probability(p) + probability(q). (Skyrms 1984: 21; emphasis in the original)

It "had better be" because the penalty for violation is a Dutch Book. As Skyrms notes, and as we saw in section 1 above, the argument is straightforwardly extended to the countably infinite case.

But more than consistency in the sense of not evaluating the same event differently is required to force P(A∨B) = P(A) + P(B). Consistency in that sense amounts only to one's evaluation being a functional; it certainly does not follow that it must be an additive one. To ensure that, one must stipulate it. Thus, to proceed from saying that P(A) is my fair betting quotient on A, and P(B) is my fair betting quotient on B, to the conclusion that the betting quotient determined by the sum of two bets at those odds is my fair betting quotient I clearly need the additional premise that I regard the sum of two fair bets as fair: i.e., one needs (*) – or at any rate the two-dimensional version of (*) – as de Finetti saw. Indeed, in a footnote Skyrms implicitly concedes the point by making an additional assumption whose consequence is that if the expected value $EV(W_1)$ of a bet W_1 is equal to $EV(W_1')$, and $EV(W_2) = EV(W_2')$, then $EV(W_1+W_2) = EV(W_1'+W_2')$, which, subject to the usual proviso about small stakes, implies (*) (Skyrms 1984: 123, n. 4).

4   Countable additivity

The foregoing does however raise questions of its own, in particular the following two:

(a) If Dutch Book arguments are redundant justifying the probability axioms (given (*)), why did de Finetti invest them with such importance?

(b) Why restrict (*) to finite sums? Isn't this just an ad hoc way of allowing non-countably additive assignments to be consistent?

I will deal with these in turn.

(a) Invulnerability to a Dutch Book is a concept of deep theoretical significance. De Finetti proved that a non-Dutch Bookable set of probability evaluations of an arbitrary set of propositions can be extended to all the propositions in any including algebra (1972: 78); it is well known that this is not true for countably additive measures. This is a suggestive result, recalling the deductive fact that a two-valued finitely additive measure (a Boolean valuation) can be extended to any algebra, and so establishes yet another link with deductive logic.

Now to (b), I remarked earlier that it is in de Finetti's discussion of the countable additivity issue that we see an appeal to logical or quasi-logical considerations trump operational, decision-theory oriented criteria. It is precisely because he sees countable additivity as importing substantive content, and thereby exceeding the remit of mere logic, that he rejected it. For example, he claimed, plausibly, that there is nothing intrinsically in-consistent in a uniform probability distribution over a countable partition, and he provides a compelling illustration: suppose you assign a uniform density distribution over [0,1], and you are then told that the true value is rational. Yet countable additivity implies that the resulting distribution over the rationals in [0,1] is heavily skewed adding, according to de Fi-netti, discriminatory information where none was supplied in the prem-ises: "Here the *content* of my judgment enters into the picture" (de Finetti 1974: 123, emphasis in the original). As such it violates a condition fa-miliar from deductive logic, that valid inferences (in this case from the null premise set) are non-ampliative:

> as with the logic of certainty, the logic of the probable adds *noth-ing* of its own: it merely helps one to see the implications of what has gone before. (de Finetti 1974: 215)

Much the same could be said about the celebrated Bayesian convergence-of-opinion theorems, which in their strong "with probability 1" formula-tion require countable additivity. Since for a countably additive distribu-tion over a countable partition some finite subset will carry almost all the probability, a typical such result implies that if a hypothesis H about a data source generating countably infinite data sequences is false the prob-ability that it will be falsified after any given finite number of observa-tions must tend to 0. It follows that sufficient positive evidence will push the probability of H arbitrarily close to 1 (Kelly 1996: 321-330). Kelly's own assessment might well have come from de Finetti himself:

> If probabilistic convergence theorems are to serve as a philosophi-cal antidote to the logical reliabilist's concerns about local under-determination and inductive demons, then countable additivity is elevated from the status of a mere technical convenience to that of

a central epistemological axiom favoring scientific realism. (Kelly 1996: 323)

De Finetti's claim that countable additivity forbids what intuitively are perfectly consistent evaluations can also be made with reference to a long-run frequency account of objective probability (de Finetti 1972: 89-90 mentions this but, presumably since he repudiates objective probability, does not stress it). Those Bayesians who are prepared, unlike de Finetti, to admit a dualistic interpretation of the probability axioms almost universally believe that one's probability evaluations should be matched to objective probabilities in default of any further discriminating information. That view has been one of the founding principles of the Bayesian theory and was used by Bayes himself in his justly celebrated derivation of the posterior distribution of a binomial parameter. To rule it out a priori certainly seems to go far beyond a consistency constraint. But its negation is a consequence of adopting countable additivity since there are finitely but not countably additive limiting relative frequency distributions. Indeed, there is a model of the de Finetti uniform distribution over a countable partition in the von Mises Collective consisting of any permutation of the set of natural numbers, with attributes the singletons $\{n\}$, $n = 0,1,2, \ldots$ The limiting relative frequency of each attribute exists and is equal to 0, though the countable union (disjunction) has of course limiting relative frequency 1 (von Mises's axiom of randomness, in Church's recursion-theoretic form, is satisfied, if rather trivially). It is well-known that events with well-defined limiting relative frequencies do not always form a field, but it follows from a result of Kadane and O'Hagan (1995) that this particular distribution can be extended to all subsets of N.

Most critics of de Finetti's views on countable additivity nevertheless see them as the weak point in an otherwise impeccable armoury. One obvious casualty is the smooth, measure-theoretic development which has become characteristic of modern mathematical probability since the publication of Kolmogorov's hugely influential treatise (1933). Countable additivity is equivalent to continuity in the sense of Kolmogorov's Axiom 5[8], on which most of modern mathematical probability relies. It ensures among other things that probability functions can be uniquely recovered (on the Borel sets) from distribution functions; for example, only if continuity is assumed does the jump at a discontinuity point a of a

---

[8] This says that if the limit of a decreasing sequence of subsets of the possibility space is the empty set then the limit of their probabilities is zero. The domain of the probability is assumed to be σ-complete.

distribution function $F^X(x)$ give the probability that X=a. De Finetti's response to this charge was, broadly speaking, that mathematics should be servant rather than master and in particular it should not be allowed to dictate first principles, or what he calls "conceptual issues" (1972: 89). But many critics argue that violating countable additivity itself contravenes intuitively compelling principles of sound reasoning. A striking example is a failure of conglomerability: non-countably additive functions on infinite algebras are nonconglomerable with respect to some countable partition $\{B_i\}$, in the sense that there is an A such that $P(A|B_i)$ lies within a closed interval I for each i, but P(A) lies outside I. De Finetti himself calls this consequence the "paradox" of nonconglomerability. It does look rather like a failure of an infinitary version of the deductive law of "or"-elimination, but de Finetti points out that the appearance of conflict is dispelled by noting that "$P(A|B_i) = p$" cannot consistently be read "If B_i then P(A) = p" (1972: 104).

There are some further objections to nonconglomerability which de Finetti does not mention, usually because he has already implicitly answered them. Thus, (i) nonconglomerable probabilities can be Dutch Booked, and also (ii) dominance with respect to some countable partition fails with nonconglomerability (there are acts f, g such that f is weakly preferred to g given B_i for all i, but g is preferred to f). The Dutch Book in question is against a countably infinite set of bets, on A conditional on each B_i, and one against A. But as we know from the earlier discussion, as an argument for incoherence this implicitly assumes the validity of countable additivity, which is well known to be equivalent both to conglomerability with respect to countable partitions, and also to countable dominance. So in each case using nonconglomerability and its consequences as an objection to finite additivity implicitly assumes what its sets out to justify.

5   Probability and logic

In the foregoing we have seen that de Finetti's view of the probability axioms was that they are mere constraints of consistency for sets of probability-evaluations, *finite and infinite*, and that for this reason the theory of probability should be seen as a branch of logic. It was this same criterion of being a mere consistency principle by which he judged – and rejected – the claim of countable additivity to axiomatic status. At any rate, what emerges from this view is a theory which in its salient features has a good deal in common with deductive logic (of being *a local, compact and non-ampliative theory of reasoning of universal scope*).

But de Finetti never developed his logical ideas in a way that embedded them within contemporary developments in formal deductive logic, in his later writing stressing the operational aspect of coherence[9]. Many people today – particularly Bayesians – still cannot see how a theory of coherent subjective probability has any common ground with post-Fregean formal logic[10]. If it is not obvious to probabilists, however, it was to at least one logician working with Carnap, Haim Gaifman. Gaifman opened his seminal paper (1964) with the observation that an unconditional subjective probability generalises a classical truth-valuation of sentences with respect to a relational structure. A relational structure for a first order language L can be represented as a pair (D,m), where m is an additive function assigning values in {0,1} to atomic sentences of L(D) (L plus enough constants to name all members of D) and extends the valuation to all sentences via a standard recursive definition. Gaifman's probability models are structures (D,m) where m is a finitely additive function taking values in [0,1].

The only disanalogy to a truth-valuation is that probabilities on atomic sentences do not in general determine values on all sentences of the language, but one of Gaifman's most important results proved in this paper was to show that adjoining a condition now called the *Gaifman Condition*, which stipulates that the probability of an existentially quantified sentence is the supremum of the probabilities over all finite disjunctions of its instances, a probability defined on the quantifier-free sentences of a first order language (and hence finitely additive) determines a unique extension to all the sentences of L. This result, a logical analogue of the extension theorem for countably additive measures on event-algebras, is like countable additivity itself tantamount to assuming that the universe is exhausted by a countable set of instances.

Gaifman's innovative work left open two questions. One is whether there is a straightforward extension to languages allowing a closer approximation to the contemporary mathematical theory of probability

---

[9] With the result that so many people now see his work as exclusively operationalist. It is these same people on the whole who also have little sympathy for his defence of finite additivity.

[10] Kyburg's reason for using "coherence" and not "consistency" to translate "cohérence" in de Finetti's original French 1937 ("consistent" *chez* logicians just means non-contradictoriness while "cohérence", in de Finetti's sense, imposes additional constraints on beliefs), is typical of this sort of view (Kyburg and Smokler 1980: 55).

which typically deals not just with finite algebras but σ-algebras (recall that de Finetti's simple countable lottery involves a probability defined on the infinite disjunction of the propositions E_i, $i = 1, 2, \dots$, to which it assigns probability 1), and whether the theory of such probability assignments could itself be formalised and a corresponding effective proof theory developed for it. A paper published shortly after Gaifman's by Scott and Krauss (1966) went some way to answering these questions. The smallest extension of first order logic to incorporate countable conjunctions and disjunctions is an infinitary language of type $L_{\omega1,\omega}$, hereafter just L, which allows closure of the class of formulas of a base first order language under the countable operations while retaining only finite strings of quantifiers[11], and Scott and Krauss defined a class of probability model-structures for such a language broadly following Gaifman. The principal difference is that in Scott and Krauss's treatment the probability function is a countably additive probability on a complete Boolean algebra B, into which the sentences of L are mapped by a valuation function which determines a σ-homomorphism from the Lindenbaum sentence algebra of L to B.

So much for the probabilistic model theory. What about a corresponding proof theory? In first order logic we don't usually associate a proof theory with proving statements assigning truth-values to object-language sentences. There are two reasons for this. One is that by Tarski's well-known theorem, no language capable of defining a certain basic class of arithmetic functions can define its own truth-predicate, and the other is that we don't need to: with only two truth-values the usual truth-table rule allows us to regard $\neg A$ as implicitly asserting that A is false, so explicit mention of the truth-values can be dropped without loss. But as Smullyan showed, any standard first order proof theory implicitly determines consequences of statements like "A is false" (and he adapts the well-known tableau method to doing calls "signed tableaux" (1968: 15-24)); Tarski's theorem is not violated because no embedding of "is true" is allowed for the "signed" sentences. We could rewrite Smullyan's "signed" formulas AT (A is true) and AF (A is false) as V(A)=1 and

---

[11] Many notions can be defined in these languages which are not first order definable, for example finiteness, and being the standard model of Peano arithmetic. But the logic is not compact (or even weakly compact), though there is a weak completeness theorem: if A is a semantical consequence of a *countable* set of formulas then it is a provable consequence (though because of the introduction rule for the countable conjunction operator proofs can now be countably infinite). For more on the properties of these languages see Keisler 1971.

V(A)=0, and thereby have an effective proof theory for assignments of *numerical values* to formal sentences. Scott and Krauss construct an analogous proof theory for a certain class of descriptions of assignments of real values in [0,1] to sentences of the infinitary language. This class comprises the sentences definable in algebraic real number theory, i.e. in the first order language M of real closed fields, famously proved complete and decidable by Tarski. These formal descriptions, called "probability assertions" by Scott and Krauss, are finite sequences $(\varphi, \mathbf{A})$ where φ is a quantifier-free formula of M[12] with n free variables and $\mathbf{A} = (A_1,...,A_n)$ are sentences of L. $(\varphi, \mathbf{A})$ "says" that the probabilities of A_1, ... , A_n satisfy the relation φ. So $(\varphi, \mathbf{A})$ is true in a model with probability function m iff $\varphi[m(A_1), ... , m(A_n)]$ is true in the real numbers. *Consequence* and *consistency* are defined analogously to their usual semantic meanings. The valid sentences, i.e. consequences of the empty set of probability assertions, are called "probability laws" by Scott and Krauss.

Scott and Krauss proceed to prove a completeness result, namely that the class probability laws is "effectively enumerable" (the reason for the scare quotes will soon be apparent). The standard way of proving such results in more familiar logics is to axiomatise the class of valid statements. Scott and Krauss do not do this but instead directly construct an algorithm for enumerating these assertions (1966: Theorem 6.7). While this is indeed remarkable, a qualification needs to be made which removes a good deal of its intuitive appeal. This is that "effectively" has to be given a much stronger sense than "recursively" in the sense of recursion theory on N[13]. Scott and Krauss's algorithm depends on an enumeration of the valid sentences of L, but the infinitary nature of the formulas and derivations in L means that, though they can be coded numerically, the coding structure has to have the cardinality of the continuum: in fact, a natural coding structure is the set of real numbers, and the class of valid sentences of L determines an analytically definable set of reals (to be precise, a Σ^1_2 set). By way of comparison, or rather contrast, the recursively enumerability of the set of valid sentences of a countable first order language, under a Gödel coding in N, means that it is very low down in the arithmetical hierarchy (i.e. the hierarchy of sets definable by formulas of

---

[12] This is no restriction since the theory of M admits quantifier-elimination, and by a famous result of Tarski is complete and decidable.

[13] If the base logical language is a first order language with at most countably many constants then the corresponding set of probability laws *is* recursively enumerable (Scott and Krauss 1966: Theorem 7.6).

first order arithmetic; the analytic hierarchy consists of sets definable by formulas of second order arithmetic, in which real analysis can be developed). What is effectively specifiable is the description of the proof system for L, but as we see that is not equivalent to saying that the theorems are effectively enumerable in anything like the sense that the theorems of pure first order logic are effectively enumerable (by a suitable Turing machine).

While extremely complex relative to that of first order logic, this proof theory for probability assertions is nevertheless still very weak in relevant respects. For example, the statement that a countable set of mutually inconsistent sentences whose disjunction is a logical truth has a probability summing to 1 cannot even be formulated as one of Scott and Krauss's probability assertions, and a fortiori is not a probability law. M itself offers a very limited medium for discussing probability assignments even where the base language is L. Much of standard mathematical probability theory quantifies over *functions* (random variables, sample functions etc.). Putting that in a formal setting means employing second languages, in particular that of second order arithmetic, whose valid sentences are not even second-order definable, and indeed are so strong that they decide some celebrated questions of set theory known to be undecidable on the basis of the standard ZFC[14] axiomatisation, among them the generalised continuum hypothesis and the existence of strongly inaccessible cardinals.

The preceding observations suggest that the enterprise inaugurated by Carnap, of trying to wed probability to logic in the form of a function defined on the formulas of a formal language, cannot aspire to anything like the expressive and deductive power of informal mathematical probability theory and at the same time have an authentically effective proof theory.

6   Finale

In the light of the foregoing I would like to end with a proposal, one much the same as I think de Finetti himself would have recommended had the issue been explicitly put to him: regard the object-languages of the probability logic as the usual algebras or σ-algebras of contemporary probability theory, with probabilistic reasoning, including reasoning with conditional probabilities, done in an informal metatheory consisting of

---

[14] "ZFC" denotes the Zermelo-Fraenkel axioms plus that of Choice.

the usual mathematics of analysis and set theory[15]. Deductive consistency and probabilistic consistency are thus subspecies of the same fundamental notion of the solvability of equations subject to constraints: those of a classical truth-valuation in the deductive case, and the rules of finitely additive formal probability in the probabilistic case.

Of course, this raises anew the question of the status of countable additivity, and for that matter the status of the finitely additive probability axioms themselves. It would be nice to have one answer to both questions. I believe there is one, in the well-known work of R.T. Cox (1961) (and independently I.J. Good 1950), which shows that a finitely additive conditional probability is a rescaling of any adequate numerical measure of belief (as is also the odds measure). Moreover, the proof of this result is limited to finitely additive measures: no natural extension of Cox's assumptions gives countable additivity. Cox's proof has been criticised in its details[16], but the objections can, I think, be answered without too much difficulty. This is not the place to discuss the matter further, and I shall simply refer the reader to the discussion in Howson 2008. That granted, I propose that *the rules of finitely additive probability are nothing else than a genuine logic of uncertain reasoning*, just as de Finetti, prescient as always, believed[17].

References

Coletti, G. and Scozzafava, R. (2002). *Probabilistic Logic in a Coherent Setting*. Dordrecht: Kluwer.

Cox, R.T. (1961). *The Algebra of Probable Inference*. Baltimore: The Johns Hopkins Press.

de Finetti, B. (1936). "La Logique de la Probabilité". In *Actes du Congrès International de Philosophie Scientifique* IV. Paris: Hermann, pp. 1-9.

de Finetti, B. (1937). "Foresight, Its Logical Laws, Its Subjective Sources". Translated and reprinted in Kyburg and Smokler (1980), pp. 53-119.

de Finetti, B. (1949). "Sull'impostazione assiomatica del calcolo delle probabilità". *Annali Triestini* XIX, pp. 29-81.

---

[15] As a nice bonus one could then claim first order set theory supplemented by Kolmogorov's axioms (without continuity) as a formalised proof theory. It would of course be incomplete, but incompleteness is endemic: in second order set theory or second order arithmetic it infects the underlying deductive logic.

[16] Halpern (1999).

[17] This appears to be also the view of the authors of a recent book in which the informal theory of coherent probabilities is presented explicitly as a logic (Coletti and Scozzafava 2002; their excellent book is strongly recommended).

de Finetti, B. (1972). *Probability, Induction and Statistics*. London: Wiley.

de Finetti, B. (1974). *Theory of Probability*, vol. 1. London: Wiley. (English translation of *Teoria delle Probabilità*, Turin: Einaudi, 1970).

Gaifman, H. (1964). "Concerning Measures in First Order Calculi". *Israel Journal of Mathematics* 2, pp. 1-18.

Galavotti, M.C. (2005). *Philosophical Introduction to Probability*. Stanford: CSLI Publications.

Gillies, D.A. (2000). *Philosophical Theories of Probability*. London: Routledge.

Good, I.J. (1950). *Probability and the Weighing of Evidence*. London: Charles Griffin.

Halpern, J.Y. (1990). "An Analysis of First-Order Logics of Probability". *Artificial Intelligence* 46, pp. 311-350.

Halpern, J.Y. (1999). "Cox's Theorem Revisited". *Journal of Artificial Intelligence Research* 11, pp. 429-435.

Howson, C. (2008). "Can Probability be Combined with Logic? Probably". *Journal of Applied Logic* (forthcoming).

Howson, C. and Urbach, P. (2006). *Scientific Reasoning: the Bayesian Approach* (third edition). Chicago: Open Court.

Kadane, J.B. and O'Hagan, A. (1995). "Using Finitely Additive Probability: Uniform Distributions on the Natural Numbers". *Journal of the American Statistical Association* 90, pp. 626-631.

Keisler, H.J. (1971). *Model Theory for Infinitary Logic*. Amsterdam: North Holland.

Kelly, K. (1996). *The Logic of Reliable Inquiry*. Cambridge: Cambridge University Press.

Kolmogorov, A.N. (1933). *Grundbegriffe der Wahrscheinlichkeitsrechnung* (English translation: *Foundations of the Theory of Probability*, New York: Chelsea, 1933).

Kyburg, H. and Smokler, H., eds. (1964, 1980). *Studies in Subjective Probability*, second edition. New York: Wiley.

Maher, P. (1993). *Betting on Theories*. Cambridge: Cambridge University Press.

Ramsey, F.P. (1926). "Truth and Probability". Page references are to *The Foundations of Mathematics*, ed. by R.B. Braithwaite. London: Kegan Paul, Trench, Trubner and Co, 1931, pp. 156-198.

Schick, F. (1986). "Dutch Bookies and Money Pumps". *The Journal of Philosophy* 83, pp. 112-119.

Scott, D. and Krauss, P. (1966). "Assigning Probabilities to Logical Formulas". In *Aspects of Inductive Logic*, ed. by J. Hintikka and P. Suppes. Amsterdam: North Holland, pp. 219-264.

Skyrms, B. (1983). "Zeno's Paradox of Measure". In *Physics, Philosophy and Psychoanalysis*, ed. by R.S. Cohen and L. Laudan. Dordrecht: Reidel, pp. 223-254.

Skyrms, B. (1984). *Pragmatics and Empiricism*. New Haven: Yale University Press.

Smullyan, R. (1968). *First-order Logic*. Berlin: Springer.

Spielman, S. (1977). "Physical Probability and Bayesian Statistics". *Synthese* 36, pp. 235-269.

Villegas, C. (1964). "On Qualitative Probability σ-Algebras". *Annals of Mathematical Statistics* 35, pp. 1787-1796.

4

De Finetti, Chance, Quantum Physics

SANDY L. ZABELL

1   Introduction

De Finetti provocatively asserts *probability does not exist*; that is, probability is a purely subjective quantity representing a judgment on our part about the world, rather than being a feature of the world, "something endowed with some kind of objective existence" (de Finetti 1974: X). Many people find such a statement – at least initially – puzzling; how can one deny the existence of objective chance when we seem to encounter it at every turn, both on the macroscopic level of our everyday experience (as in Pearson's "the chances of death"), and also – or so quantum physics seems to tell us – on the microscopic level of atoms, molecules, and subatomic particles?

Understanding the role and meaning of probability in quantum physics seems particularly daunting; unlike the probability of a coin coming up heads landing on red in roulette, or getting a straight flush in Texas hold 'em, all things we have direct experience of and intuitions about, the phenomena of quantum physics seem counter-intuitive, remote from everyday experience, and require sophisticated mathematical tools for their description.

De Finetti found the subject of quantum physics fascinating and challenging; so challenging in fact that he did not discuss it in detail until relatively late in his career, in Sections 9-12 of the *Appendix* to his magnum opus *Teoria delle Probabilità* (de Finetti 1970). The present paper has its origins in some sense in a conversation many years ago with Richard Jeffrey, who remarked that it would be interesting to understand de Finetti's extended remarks on quantum mechanics in the *Teoria*. The cur-

Bruno de Finetti, Radical Probabilist
Maria Carla Galavotti (ed.)
Copyright © 2009

rent essay is an attempt in that direction, an analysis of de Finetti's thoughts on objective chance and quantum mechanics, placing them in the context of his predecessors and successors.

2 "Probability does not exist"

In order to have a solid foundation for a subjective theory of probability, one needs both an operational definition of the concept, and a persuasive derivation of its properties. In the years immediately after World War I, Frank Ramsey in England and Bruno de Finetti in Italy independently achieved this. The details of their systems differed: Ramsey gave a simultaneous derivation of probability and utility based on a system of consistent preferences (see, e.g., Zabell 1991); de Finetti one based on betting odds and the use of the Dutch book argument.

But in addition to his constructive analysis, de Finetti also rejected outright the notion of objective probability or chance. It is precisely this stance that makes de Finetti "radical", for not all subjectivists feel the need to argue this. Ramsey, for example, in his essay "Truth and Probability" (Ramsey 1931: 158-9) starts out, "in the hope of avoiding some purely verbal controversies", by (albeit "temporarily") making the admission that the frequency interpretation of probability may be the one of greatest value to science (even though, Ramsey adds, this "does not express my real opinion" and "I myself do not believe this"). Establishing the first position (subjective probability exists) does not require abandoning or actively discrediting the second (an objective theory of probability). There are in fact many prominent Bayesians who subscribe to some form of physical probability or objective chance, including I. J. Good (1965: 6), David Lewis (1980), and Richard Jeffrey (1986: 10).

De Finetti's "radical probabilism" strikes many as difficult to understand, almost paradoxical, so it is interesting to note that de Finetti himself did not come to such a position easily. He tells us:

> I would like to add that I understand very well the difficulties that those who have been brought up on the objectivistic conceptions meet in escaping from them. I understand it because I myself was perplexed for quite a while some time ago (even though I was free from the worst impediment, never having had occasion to submit to a ready made and presented point of view, but only coming across a number of them while studying various books and works on my own behalf). It was only after having analyzed and mulled

over the objectivistic conceptions in all possible ways that I arrived, instead, at the firm conviction that they were all irredeemably illusory. It was only after having gone over the finer details, and developed, to an extent, the subjectivistic conception, assuring myself that it accounted (in fact, in a perfect and more natural way) for everything that is usually accredited, over-hastily, to the fruit of the objectivistic conception, it was only after this difficult and deep work, that I convinced myself, and everything became clear to me. (de Finetti 1974: 4-5)

So how does one arrive at such a conclusion? There are both destructive and constructive aspects. On the destructive side one must expose the logical shortcomings of the objectivist account; on the constructive side one must "save the phenomena"; that is, account for in a purely subjective manner the phenomena "explained" by objective chance.

3 Objective chance

By April 1928 de Finetti had "prepared a complete exposition of the foundations of probability theory according to my point of view" (de Finetti 1931: 221). Encountering, however, not only unexpected resistance to his ideas but sometimes even outright misunderstanding, de Finetti realized that a more detailed presentation was necessary. The result was a series of papers, one of which, "Probabilismo" (de Finetti 1931) contains a careful and systematic philosophical discussion of his views.

"Probabilismo", arguing at length the philosophical basis for de Finetti's total rejection of objectivism, may well retrace the path he took in "this difficult and deep work". De Finetti argues his position here with great force, clarity, and depth, although his thesis is so radical that it is hard to grasp it in its entirety on a first reading. (One is reminded of Wittgenstein's comment at the beginning of his *Tractatus*: "Perhaps this book will only be understood by someone who has himself already had the thoughts that are expressed in it".)

"Probabilismo" deserves close and careful study. Although de Finetti's later paper, "La prévision, ses lois logiques, ses sources subjectifs" (de Finetti 1937) is far better known, de Finetti tells us at its beginning that "Probabilismo" is in fact a "more complete statement of my point of view, in the form of a purely critical and philosophical essay, without formulas" (de Finetti 1937: fn. 1). In particular, Sections 5-20 of "Probabilismo" contains perhaps de Finetti's most sustained attack on the concept of objective probability. Understanding in detail de Finetti's total rejection of objective chance will help us later in understanding his view of the role of probability in quantum physics.

There are really two basic arguments for the existence of objective chance: physical symmetry and the perceived stability of long-term frequencies. These are, in many ways, complementary: the first entails a single-case, dispositional, a priori concept; the second a frequency-based, empirical, a posteriori one.

3.1   Physical symmetry

Consider the case of an urn having an equal number of red and black balls. If we select a ball "at random", that is, regard every ball as being equally likely to be chosen, then coherence requires we assign a probability of 1/2 to the event the ball is red. This number is based on both *objective facts* (the composition of the urn), and *personal judgment* (the balls in the urn are thought to be equally likely to be chosen); it is therefore necessarily subjective in nature – a judgment by us about an event – and not objective, in the sense that it is a property of some aspect of the world. Such initial judgments of equality are often revised in the light of experience. Real dice are never exactly "fair", witness Weldon's dice data (Kemp and Kemp 1991); and true random sampling is rarely present in nature or human affairs. (Statistical sampling practitioners in fact often find they must go to great lengths to achieve it; see Fienberg 1971, for one particularly interesting example).

The classical theory of probability is based on the assumption that outcomes form a set of equally probable (or "equipossible") cases; more complex forms of observation are then sometimes likened to drawings from the "urn of nature". (Price's discussion of inductive inference in his Appendix to Bayes's famous essay is an early and striking example of this; see Zabell 1997). But just what does it mean to say that the cases are "equal"? Hacking (1971) argues that in the infancy of probability theory mathematicians such as James Bernoulli had something like an early form of propensity theory in mind. De Finetti (1931a: §6-9) argues that in every case an appeal to equally probable cases inevitably involves an irrefragable element of subjective judgment. It is "absurd, even if it were not meaningless, to consider probability as a mysterious and unreachable metaphysical entity, existing in abstraction, on which the occurrence of an event somehow or other depends" (§8, 178). Saying that the cases only differ in respect of causal circumstances having no influence on the outcome is futile:

> But let us examine our conscience, and see when it is that we admit that a circumstance can influence a certain event. Isn't it precisely when knowledge of it influences our probability judgment? Do we mean anything more than that? No matter what we say or think, in the end we come to this: the concept of cause is only subjective, and it depends essentially on the concept of probability. (§9, 179)

Here we see very clearly Hume's views echoed in de Finetti (cf. de Finetti 1937: 155; 1972: 183).

De Finetti does make an interesting concession (§16, 190): "as far as dice and lotteries are concerned, I can admit that it is of little practical interest whether the notion of equally probable cases has an objective meaning or whether its meaning is subjective but corresponds to a state of mind so natural as to seem universal". So why bother? The reason: it is all too easy to fall into error "if we depart, even a little, from the artificial and schematic examples that conceal every conceptual difficulty". (Here de Finetti oversteps himself in his zeal, suggesting that the equally probable setting includes only trite instances devoid of genuine interest; in reality it encompasses the entire theory of sampling from a finite population.)

But what are examples that are not "artificial and schematic"? This brings us to our second category.

3.2   Frequency

Just as the use of symmetry arguments invokes the concept of "equally probable cases", the frequency approach invokes "trials of the same phenomenon" (§10, 181). Here too the subjective element is apparent: "Even on first analysis, it will turn out that this has only a conventional meaning, case by case". De Finetti's verdict is harsh: the concept is "arbitrary", it is "vain to look for a philosophical substratum", vain to think of it as "something meaningful in itself".

Nevertheless, the frequency interpretation always has a siren-like attraction. De Finetti quotes Poincaré (§19, 194): insurance companies "distribute to their shareholders dividends whose objective reality is incontestable". Statistical regularities abound, surely this evidence of the objective nature of probability.

All this is merely a confusion of thought. "I can foresee frequencies almost with certainty. This is an objective fact." But it is not evidence that vindicates the frequency theory as a manifestation of some supervening objective reality independent of our judgment; evidence that probability is, as Maurice Fréchet asserts, "a purely experimental notion, the notion of a physical constant whose measure we can approach by making actual observation of the frequency of a fact" (§14, 186). Suppose, for example, one selects not one, but an entire sequence of balls from our urn (either replacing the chosen ball each time, or not, the mathematics is the same), and suppose further we regard all possible sequences as being equally likely to occur. Then for sufficiently long sequences it follows *on purely mathematical grounds* that one *expects* the fraction of red balls in a sufficiently long sequence to reflect the contents of the urn.

To see this, suppose the fraction of reds in the urn is p, consider sequences of length n, and let A_δ denote the set of sequences having their fraction of reds within δ of p. If the sequences are long enough (if n is sufficiently large), then most (all but a fraction ε) of them will lie in A_δ (this is a combinatorial version the law of large numbers); because the sequence are judged to be equally likely, it follows that the probability of A_δ must be judged to be close to one (greater than $1 - \varepsilon$). This probability arises once again from a combination of objective facts and personal judgment, but is not in itself an objective property of some aspect of the world. If the fraction of red balls in the sequence one observes is approximately the same as the fraction of red balls in the urn, one should not be surprised: *this is precisely what we expected to see.* If one thinks an event is "practically certain" to occur, it is fatuous to be impressed if it in fact occurs.

Popular belief in the so-called "law of averages" expresses a related and widespread misapprehension that somehow there is a mysterious driving force maintaining the stability of long-term frequencies. Edgar Allan Poe (in the last paragraph of his famous short story "The Mystery of Marie Roget") is merely the best known out of a vast multitude who mistakenly think an observed excess of heads in a sequence of fair coin tosses makes it *less* likely that the next toss will come up heads!

4 The de Finetti representation theorem

One of the challenges for the foundations of both probability and statistics is that statisticians often use one of two very different paradigms in applied statistics: sampling without replacement from a finite population, and independent trials of the same phenomenon (it is no accident that these correspond to the two versions of objective chance discussed above). Attempts to include one paradigm within the other seem forced: viewing either independent trials as sampling without replacement from a supposed "infinite hypothetical population" (R. A. Fisher); or viewing a sample drawn without replacement from a finite population as a single realization in an infinite sequence of possible samples. De Finetti realized that the concept of *exchangeability* furnished both a natural unified framework for the two models, and a natural account of the inductive process.

4.1   Exchangeability

Suppose one observes a sequence of outcomes, $X_1, \ldots, X_n$, each belonging to one of t distinct categories $c_1, \ldots, c_t$ (for example, different colors or species). To be concrete, we might think of the observed outcomes as being the letters in some alphabet containing t letters. A probability assignment on such sequences is said to be *exchangeable* if every n-long sequence of letters having the same number of letters of each type is assigned the same probability. For example, if $n = 12$, then under exchangeability the three sequences

AAAABBBBCCCC, ABCABCABCABC, ABCBACCBCAAB

are assigned the same probability.

A word here about notation and terminology may be helpful. In the calculus it is a useful abuse of notation to use the symbol $f(x)$ to refer to both a function (the function f) and a number (the value $y = f(x)$ the function f assigns to the number x). Similarly, it is sometimes useful to employ the notation $X_1, \ldots, X_n$ to represent both a sequence of "random variables" (that is, functions on a given sample space), and sometimes a specific (but unknown) sequence. The first is current standard mathematical usage, the second much closer to that of de Finetti (who deprecated the first as being a Procrustean bed). In either case, one says the sequence is exchangeable if there is an understood or implicit exchangeable probability assignment P on its possible realizations.

If n_i denotes the number of times the i-th letter appears in an exchangeable sequence $X_1, \ldots, X_n$, then specifying the probability assignment P for the sequence reduces to finding $P(n_1, \ldots, n_t)$, the probability of the frequency counts $n_1, \ldots, n_t$. For example, there are 34,650 different ways one can arrange four A's, four B's, and four C's, and the probability of each of these arrangements given exchangeability is $P(n_A, n_B, n_C)/34,650$. In statistical parlance, one says the counts $n_1, \ldots, n_t$ form a set of *sufficient statistics* for the sequence.

4.2 The finite representation theorem

Given a set of counts $n = (n_1, \ldots, n_t)$, imagine an urn containing n_1 balls of each type, and suppose one successively draws out "at random" *without replacement* each ball in the urn ("at random" meaning that all possible sequences are judged equally likely). There are a total of $(n_1 + \ldots + n_t)!/(n_1! \ldots n_t!)$ such sequences; the (exchangeable) probability assignment H_n according each of these equal weight is called the *hypergeometric distribution*. If, more generally, $X_1, \ldots, X_n$ is any exchangeable sequence whatsoever, and $P(n)$ the corresponding probability assignment on the counts n, then the overall probability assignment P on the set of sequences is a mixture of the hypergeometric probabilities H_n using the weights $P(n)$; compactly this can be expressed as $P = \Sigma_n P(n) H_n$.

This result is the *de Finetti representation theorem* for a *finite* exchangeable sequence; although not quite as well known (or appreciated) as its big brother, the representation theorem for an *infinite* exchangeable sequence, it would be a serious mistake to underestimate it. To begin, thanks to the representation, there is a drastic reduction in the number of independent probabilities to be specified; in the case of tossing a coin 10 times, for example, from $2^{10} - 1 = 1023$ to 11. (Note also that the finite representation theorem makes no reference whatsoever to the infinite, in either its statement or assumptions.)

4.3 The infinite representation theorem

It is an interesting mathematical fact that some finite exchangeable sequences are *extendible*, and others are not. (The sequence $X_1, \ldots, X_n$ is *r-extendible* if it is the initial segment of a longer exchangeable sequence $X_1, \ldots, X_r$, for some $r > n$.) This phenomenon should not be thought of as being somehow pathological: for example, none of the hypergeometric assignments H_n are extendible. (Intuitively this is not surprising: one runs out of balls to draw from the urn!) On the other hand, if $r = (r_1, \ldots, r_t)$ and $r = r_1 + \ldots + r_t > n$, then restricting $H(r)$ to $X_1, \ldots, X_n$ is a simple means of exhibiting an extendible exchangeable probability assignment (one is drawing n balls at random without replacement from an urn with $r > n$ balls).

Suppose the exchangeable sequence $X_1, \ldots, X_n$ is *infinitely* extendible; that is, it is r-extendible for all $r > n$. If $r = (r_1, \ldots, r_t)$ are the counts corresponding to an extension $X_1, \ldots, X_r$, then $P(r)$ can be viewed as a probability not just on r, but also on the scaled frequencies $(r_1/r, \ldots, r_t/r)$. Denote this probability as $Q(r_1/r, \ldots, r_t/r)$. The fractions $r_1/r, \ldots, r_t/r$ are themselves a probability on the set $\{1, \ldots, t\}$, and so $Q(r_1/r, \ldots, r_t/r)$ can in turn be interpreted as a probability on Δ_t, the set of all probabilities on $\{1, \ldots, t\}$. De Finetti's justly celebrated infinite representation theorem then tells us (see, e. g., de Finetti 1937, Cifarelli and Regazzini 1996) that (a) as $r \to \infty$, the probabilities $Q(r_1/r, \ldots, r_t/r)$ converge to a probability Q on Δ_t, and (b) the probability $P(n_1, \ldots, n_t)$ can be expressed as a Q-mixture of the *multinomial probabilities* $n!/(n_1! \ldots n_t!) \, p_1^{n_1} \, p_1^{n_2} \ldots p_1^{n_t}$.

There are some nuances here. Note that there has been no mention at all of a single *infinite* exchangeable sequence $X_1, X_2, X_3, \ldots$, only the assumption that the initial finite sequence of length n can be extended an arbitrary finite length r; an Aristotelian might refer to this as being the difference between the potentially infinite and the actually infinite. Note also that nothing has yet been said about how the $Q(r_1/r, \ldots, r_t/r)$ tend to Q; in particular, whether the convergence is "in probability" or "almost surely", or something else. This last point turns out to be of considerable importance and interest from a foundational point of view.

5   The interpretation of the representation theorem

De Finetti is notable for his rejection on principled grounds of the otherwise nearly universal adoption of the axiom of countable additivity for probability measures. The Dutch book argument led in a natural way to finite additivity, but not countable additivity, and – true to his principles – de Finetti refused to adopt this extended form of the axiom. (For a survey of the application of the Dutch book argument, see Armendt 1993.)

To explore the consequences of rejecting countable additivity in the case of the representation theorem, let us consider it in its simplest form, when there are only two categories ($t = 2$), say head and tail. In this case Q can be thought of as a probability measure on the unit interval (because $p_1 + p_2 = 1$, it is enough to specify $p = p_1$). Suppose S_n records the number of heads in n trials. In this case the representation theorem takes the form

$$P(S_n = k) = \int_0^1 \binom{n}{k} p^k (1-p)^{n-k} dQ(p)$$

That is, for every $n \geq 1$ and $0 \leq k \leq n$ the probability $P(S_n = k)$ can be thought of as a mixture of binomial probabilities (the integrands) using the measure Q. The result is the same as if one adopted Q as a prior distribution for a constant but unknown chance p of heads, but is conceptually very different, because the use of exchangeability completely eliminates any need to refer to chance.

5.1 The "naïve" subjectivist

Nevertheless, for the naïve subjectivist (including at one time this author), someone, that is, who is comfortable working with the fiction of an infinite exchangeable sequence $X_1, X_2, X_3, \ldots$ of heads and tails, and who is also willing to make the assumption of countable additivity, the de Finetti representation theorem has an interesting interpretation. Such an exchangeable sequence is a special case of a stationary sequence of random variables, and it is an immediate consequence of the Birkhoff ergodic theorem that the proportion of heads S_n/n has an almost sure limit; that is, the limit

$$Z := \lim_{n \to \infty} \frac{S_n}{n}$$

exists with probability one. Moreover, the distribution of the limit Z is none other than the mixing measure Q in the de Finetti representation.

Thus such a subjectivist *must* believe in the existence of limiting frequencies (being an immediate mathematical consequence of exchangeability); and the measure Q thereby acquires a (potential) subjective interpretation: the degree of belief about the possible values of the limiting frequency (rather than Q being a purely formal mathematical object). But this does not mean such a person has now been converted to objectivism, for the question still remains: do such limiting frequencies have a right to be regarded as objective or physical properties?

The issue is: a property of what? Some philosophers regard objective chances as *propensities* ascribed to a *chance set-up* (see, e.g., Hacking 1965); for example, in the case of coin tossing, the coin *and* the manner in which it is tossed. But then one could argue as follows: if the coin were indeed tossed in an identical manner on every trial, it would always come up heads or always come up tails; it is precisely because the manner in which the coin is tossed on each trial is *not* identical that the coin can come up both ways. (Note that at this point the "naïve subjectivist" has also become a "causal subjectivist".) The suggested "chance set-up" is in fact nothing other than a sequence of objectively differing trials, which we are subjectively unable to distinguish between. This type of argument holds in general as long as we remain in the classical deterministic world; to what extent it holds in the quantum case will be discussed later.

5.2   De Finetti's interpretation of Q

De Finetti, however, views matters very differently. He showed, using just finite additivity, that the probabilities $Q(r_1/r, ..., r_t/r)$ converge "in distribution" to a (finitely additive) probability Q, and that an integral representation holds using Q. The objection to crediting such limiting frequencies as meaningful entities, however, even in the limited way discussed earlier (the "naïve subjectivist"), is twofold, on both philosophical and mathematical grounds.

First, on philosophical grounds, the use of a quantity whose value depends on the full infinite sequence is indefensible (see, e.g., de Finetti 1979: 130, cf. Jeffrey 1977). The assertion $Z = 0.5$, for example, is not a true event (that is say, something that is verifiable or can be observed), and probabilities can only be assigned to events. (The infinite sequence is a fiction, and although it might be a useful fiction, one has to be careful in its use.)

Second, on mathematical grounds, a family of consistent finite-dimensional probability distributions does not have a unique finitely additive extension to "infinitary" events such as those describing the almost sure limit behavior of a sequence (see, e.g., Cifarelli and Regazzini 1996: 258-259); thus the use of countable additivity to determine the probabilities of such events represents merely an arbitrary choice in passing from the finite to the infinite, and therefore cannot represent an aspect of physical or objective reality.

Similar issues can be seen in the use (or misuse) of statistical independence. Consider a pair of events A and B for which $P(A)$ and $P(B)$ are given, and one is interested in the probability $P(A \cap B)$. In general, because of the non-negativity and sub-additivity of P, all one can say is that $P(A \cap B)$ must satisfy the two inequalities $0 \leq P(A \cap B) \leq min\{P(A), P(B)\}$. True, if the events A and B are judged to be *independent*, then one has $P(A \cap B) = P(A) P(B)$. But adopting independence just because it provides a simple mathematical means of extending P to $A \cap B$ is obviously inappropriate; one must have some serious affirmative grounds other than mere mathematical convenience.

Two quotations from de Finetti, one early on in his career and one much later, give insight into his position. The first is from "Probabilismo":

> [T]he probability that the frequency is in an assigned interval (ξ_1, ξ_2) tends to a well determined limit. On the ordinary view, this limit is the probability of the hypothesis that the constant but unknown probability lies in the assigned interval (ξ_1, ξ_2). Then in this case we have a limit law which is the one that in the old jargon would be called "the law of the limit of the probability of the probability".
>
> But it remains the case: first, that the ordinary method is conceptually meaningless; second, that the existence of the limit law on which it is based is an extremely delicate question, and it is not advisable to take it as a starting point, even though we were to change the statement of it into a meaningful expression. (de Finetti 1931a: §22, 201)

The second is from a paper with Savage written in 1962:

> [E]verything is as if the limit distribution could be interpreted in the usual terminology as the distribution of the unknown probability conditional on the value of which the trials are independent. In somewhat less fictitious terminology, one might say that the limit distribution is the distribution of the limit frequency. This usage rests on the strong law of large numbers in the form based on complete additivity. But it is possible, as we have seen, and to my mind preferable, to stick to the firm and unexceptionable interpretation that the limit distribution is merely the asymptotic expression of frequencies in a large, but finite, number of trials. (de Finetti 1972: 216)

Note de Finetti's careful language: it "not advisable" to take the limiting frequency Z as a starting point; and it is "preferable" not to interpret the limit distribution Q as the "distribution of the limit frequency" Z. (I take de Finetti's remark, "even though we were to change the statement of it into a meaningful expression", to refer to precisely the earlier analysis of the naïve subjectivist.) But doing so is not said to be "conceptually meaningless". This may be because, although such an argument appeals to countable additivity, it is still based on a coherent probability judgment and therefore not to be cast out into the darkness. It is simply that Q, being merely one of many possible extensions of the finite dimensional distributions, is not forced on us by the constraint of coherence, and therefore it is inadvisable to base our view of inference on it.

In sum: de Finetti's infinite representation theorem provides the subjective explanation for the classical Bayesian procedure of using prior distributions on constant and unknown probabilities ("chances"). (Carnapians might say the first is the "explicatum", the second the "explicandum".) But the mixing measure is not a distribution on chances – because these are fictions – and it is preferable not to interpret Q as being a distribution on limiting frequencies. Chance (even if you could make sense of it) is unnecessary.

6   Quantum mechanics

[De Finetti] claims that "Probability does not exist" ... . By this he means that it does not exist in an objective sense, in other words he denies the existence of physical probability. Although I agree that physical probability cannot be measured without using subjective probability, I feel that to deny its existence is too extreme. It could have been consistently maintained that the probabilities underlying classical statistical mechanics are necessarily subjective, and arise because of our ignorance of the precise initial conditions, but the probabilities of quantum mechanics might well be an irreducible feature of the interaction between a physical system and a piece of physical apparatus. (Good 1977: 94)

I. J. Good raises a key issue. Even if one denies the existence of chance in the deterministic setting of classical physics, doesn't quantum mechanics force it on us in its own realm? In terms of our coin-tossing example, it is as if when we toss the coin exactly the same way each time, sometimes it comes up heads and sometimes tails. The objection of a causal subjectivist to objective chance fails. This is why de Finetti's views on quantum mechanics are of such great interest.

As noted earlier, de Finetti's most detailed discussion of quantum mechanics appears in the Appendix at the end of his two-volume magnum opus *Theory of Probability* (de Finetti 1975, Sections 9-12: 302-325). Although at first blush quantum mechanics appears to be merely one of a number of important topics whose subtleties are addressed in greater detail in the Appendix than in the text proper, upon closer examination quantum mechanics is really seen to be one of the primary motivations for writing the Appendix. (Page references in this section are to de Finetti 1975, unless otherwise stated.)

De Finetti begins the Appendix by stating his preferred formulation for probability: for *events* (rather than subsets of a given, pre-specified sample space), for considering all *coherent probabilities* on those events, extendible in principle to all events (rather than just a single probability function), and for *previsions* (expectations) on *random quantities* (random variables) satisfying the inequalities entailing coherence (rather than starting first with measure and then deriving expectation).

De Finetti acknowledges that there are subtleties associated with the concept of an "event". Above all, an event must be *verifiable*; "a notion which is often vague and elusive" (p. 260). Nevertheless, "verifiability is the essential characteristic of the definition of an event", an "unverifiable event" is an oxymoron. But there are complexities, because "there are various degrees and shades of meaning attached to the notion" (including precision, time, cost, and effort).

It is at this point that quantum mechanics is revealed as one of the principle motivations for the Appendix:

> The most precise and important [issue], however, is that which arises in theoretical physics in connection with *observability* and *complementarity*. It seems strange that a question of such overwhelming interest, both conceptually and practically ... should be considered, by and large, only by physicists and philosophers, whereas it is virtually ignored in treatments of the calculus of probability. We agree that it is a new element, whose introduction upsets the existing framework, making it something of a hybrid.

> It is our intention, therefore, to attempt to provide in this Appendix an integrated view of questions of this kind which arise in connection with events. ... [O]ur attempt will be mainly concerned with the case of theoretical physics, and will consist of little more than a comparison of the positions adopted by various other authors, plus an indication of which position seems to us to be less open to criticism ... (pp. 260-261)

Thus for de Finetti the primary issue raised by quantum mechanics is not that it entails some new form of probability, but its implications for the concept of event. Let us see why.

6.1   Early de Finetti

De Finetti tells us that his views regarding the "notion of indeterminism and the related notion of complementarity" in quantum mechanics had evolved over time. His initial position was straightforward:

> My attitude had previously consisted in rejecting *ad hoc* interpretations in relation to quantum physics in order to reduce everything, essentially, to familiar situations (to facts which were 'complementary' in the sense that they were conditional on mutually exclusive experiments; like the behavior of an object in two different destructive testing situations; or the victory of a tennis player in two different tournaments taking place at the same time in two different countries). (p. 303)

De Finetti states that this "solution seems to coincide" with B. O. Koopman's approach (Koopman 1957), but de Finetti never discussed this initial view at any length: "A mention of this can perhaps only be found in the CIME course given in Varenna in 1959". Here in fact is the totality of what de Finetti says in his CIME course:

> In the very definition of event, we have said that it ought to be a definite verifiable fact … Perhaps many events, each one verifiable, might not be simultaneously verifiable, which would lead to complications typical of quantum theory. It seems to me improper, however, to regard cases of this kind as unique to quantum theory or, still worse, as symptoms of paradoxes characteristic of the mode of thinking required by quantum theory. We can test a given object to find out at what temperature it melts, or under what pressure it splits, or how long it takes to wear out to a given extent under normal use, but we evidently cannot carry out more than one of these tests. If we express the situation saying that it is meaningful to ask whether the object is refractory, or robust, or long wearing, but not to ask two or more of these questions together, this mode of speech makes the situation seem more or less paradoxical. It is thus, only with reference to concepts for which it would seem more "natural" to expect an answer to every composite question that the typical cases of quantum theory seem paradoxical, though they have nothing in them essentially different from the first example. Similar ideas and examples are presented by Koopman (1957). (de Finetti 1972: 208)

This is indeed the same position as Koopman's. (Koopman had two decades earlier developed a theory of qualitative subjective probability, and
was someone whose views de Finetti respected.) Noting a common view
that "the classical laws of probability cease, in some sense, to be valid" in
quantum mechanics, Koopman (1957: 97) argued that "all that is needed
is to make quite clear and explicit the concept of *event*". Pointing to the
existence of "incompatible" (that is, complementary) events as the source
of the supposed problem, Koopman's solution is simple: probabilities
assignments are only meaningful for conjunctions and disjunctions of
compatible events, and this is true whether or not one is dealing with
classical or quantum events.

6.2   Intermezzo

Two years later Suppes (1961: 386-387) criticized Koopman's position,
noting a fundamental difference between the classical and quantum situations. In the example posed by Koopman (treating a rat with two different
poisons), in principle one might be able to find something in the physiol-
ogy of the rat that would enable one to determine the outcome of both
treatments. In the language of quantum mechanics, there might be an ex-
planatory "hidden variable". The standard interpretation of quantum mechanics, however, asserts that at the quantum level such hidden variables
do not exist for certain types of events.

Suppes presented his paper in September, 1959 at a conference in
Paris on "Axiomatic Method in Classical and Modern Mechanics", held
at the Institut Henri Poincaré. Eight months later Suppes returned to Paris
and met de Finetti at another conference, on decision theory, but their
conversations then appear to have only concerned the axiom of choice
and the use of countable additivity (Suppes and Zanotti 1989: 323). Suppes tells us, however, that after this first meeting,

> Over the years I had the opportunity to meet de Finetti on various
> occasions. I recall a memorable walk with de Finetti and Jimmy
> Savage in the gardens of the Villa Frascati near Rome later in the
> 1960s. On this occasion we had a long discussion of determinism
> and quantum mechanics, and what the existence of indeterminism
> and quantum mechanics implied for subjective theories of prob-
> ability. (Suppes and Zanotti 1989: 323)

This raises the intriguing possibility that de Finetti may have been led to
revisit the subject of quantum mechanics at least in part because of Sup-

pes (either because he had read Suppes 1961, or because of conversations with Suppes). In any event it is clear that during the 1960s de Finetti was once again interested in quantum physics.

6.3   Later de Finetti

During this later period de Finetti read several important books and papers on the foundations of quantum mechanics; in the Appendix he specifically cites three books: Reichenbach's *Philosophical Foundations of Quantum Mechanics* (1944), von Neumann's *Mathematical Foundations of Quantum Mechanics* (1955), and Bodiou's *Théorie dialectique des probabilities* (1964), as well von Neumann and Birkhoff's early paper "The logic of quantum mechanics" (1936). (Bodiou's book demonstrates that de Finetti's interest in the logic of quantum physics continued on into at least the middle of the decade.)

The result of all this reading was that de Finetti was led to refine rather than alter his earlier views:

> Subsequent reflection (after a good deal of reading – the most relevant being that mentioned above), has not changed my original view, but rather made it more precise.

> The solution that I will put forward is a different one [from the authors cited] but it could, in a certain sense, be seen as a simplified version of those set out by these authors. (p. 303)

De Finetti was drawn by the fact that interpreting the results of the quantum mechanics involves a new logical calculus defined on a non-Boolean lattice of propositions ("quantum logic"), distinguished from classical Boolean logic and characterized by the phenomenon of non-modularity. Here de Finetti (p. 305) identifies a subdivision: some students, such as von Neumann, see the attendant logical problems of complementarity as only arising in the quantum world, while others, such as Bodiou, see them as much more general. ("[T]he quantum calculus is simply a special case, imposed by necessity, of a general calculus of probability which we call *dialectic*. This latter, far from being an unnatural growth on the body of the classical calculus, in fact subsumes it.")

But while von Neumann and Bodiou are the technical sources of de Finetti's discussion, it is to Reichenbach that he turns for philosophical analysis. This is because, while Reichenbach refers to quantum mechanics, it is in a form that, "from a conceptual point of view, can be adapted to any context whatsoever", something that de Finetti for obvious reasons finds attractive. (Indeed, de Finetti casts his own analysis as being, in effect, a gloss on Reichenbach's, "putting forward remarks of our own as comments on his".)

For de Finetti the object is to find suitable "logical constructions" to resolve the apparent difficulties. (Note: *logical*, not probabilistic constructions.) Despite the fact that "[t]his goal does not appear to have been achieved", de Finetti believes the "correct path" to it to be "simple and straightforward". Why then is there such confusion and apparent complexity? De Finetti takes what is in some sense a Kuhnian view: there is an "initial state of confusion" arising from a novel perspective different from "one's accustomed way of seeing things", and he points to the conceptual revolutions of Copernicus and non-Euclidean geometry.

The key is to realize that "in order to find something suitable for our purpose we must bring into the world a new logic" (p. 306). This requires either (Reichenbach) introducing a third truth value ("indeterminate"), and then defining the logical operations from the resulting truth tables, or (Bodiou) introducing the logical operations directly via axiomatic definition. De Finetti discusses this new logic on pp. 306-309 and its relationship with pairs of complementarity events (that is, at least one of which must always be indeterminate) on pp. 309-313 and 322-325. It is at this point that the issue of indeterminism takes center stage.

7   The renunciation of determinism

> We ought then to consider the present state of the universe as the effect of its previous state and as the cause of that which is to follow. An intelligence that, at a given instant, could comprehend all the forces by which nature is animated and the respective situation of the beings that make it up, if moreover it were vast enough to submit these data to analysis, would encompass in the same formula the movements of the greatest bodies of the universe and those of the lightest atoms. For such an intelligence nothing would be uncertain, and the future, like the past, would be open to its eyes. (Laplace 1814: 2)

The key to understanding de Finetti's view of quantum physics is his attitude towards indeterminism. Far from questioning the claim of indeterminism in quantum physics, de Finetti argues (p. 323) that indeterminism is a more general phenomenon. "What is the difference, then, from a logical point of view, between the complementarity or non-complementarity of measurements in the case of a physicist and in that of the tailor …? [Or] that of the coin whose next toss could be made by either Peter or John…?". Reichenbach bases himself "upon an absolutely rigid division between the indeterminacy of the quantum world and the determinacy of the macroscopic world"; de Finetti rejects such a division. De Finetti quotes Reichenbach: after Peter tosses the coin,

> For us the truth value of John's statement will always remain unknown; but it is not *indeterminate*, since it is possible in principle to determine it, and only lack of technical abilities prevents us.

De Finetti rejects Reichenbach's position on a number of grounds. The first is essentially an appeal to the existence of chaotic phenomena: one cannot predict "those facts for which numerous microscopic circumstances might prove decisive". As Heraclitus famously observed, one cannot step into the same river twice.

The second is to question how the outcome of something that did not happen could in any sense be "determined":

> How could it come about that the state of the muscles, and so on, could inform us about the result of the toss that has not taken place (and why not the text of a conversation that has not taken place; the adventures of a journey not undertaken; etc.), rather than informing us directly that it is predetermined that the toss, or the conversation, or the journey, will not take place (or did not take place)?

(To take a simple concrete example in the spirit of those de Finetti suggests, the English classical scholar E. R. Dodds tells in his autobiography (1977: 79-82), of agonizing for several months over which of two women to marry, both of whom he thought he loved. In what sense could the happiness or unhappiness of both marriages be thought to be "determined" beforehand? In general, attribution of causal effect is complicated when each unit can receive only one treatment, and units are inherently unique; see, for example, the discussion of the "Rubin causal model" in Holland 1986.) In the end de Finetti concludes (p. 324):

> In my opinion, the attachment to determinism as an *exigency of thought* is now incomprehensible.

De Finetti is willing to accept from a "psychological-aesthetic angle" that someone might regard "probabilistic quantum theory as merely a partial explanation, unsatisfactory and provisional, and requiring replacing sooner or later by something deterministic"; this is in fact the Bohmian view of quantum mechanics. But de Finetti regards this (p. 325) as a "distant prospect: the foundations of physics are those we have today"; and in any case

> I think it unlikely that [the foundations of physics] can be interpreted (or adapted) in deterministic versions, like those that are apparently yearned for by people who invoke the possible existence of 'hidden parameters', or similar devices. I hold this view not only because von Neumann's arguments against such an idea seem to me convincing ... but also because I see no reason to yearn for such a thing, or to value it – apart from an anachronistic and nostalgic prejudice in favour of the scientific fashion of the nineteenth century. If anything, I find it, on the contrary, distasteful ...

> In any case, for what concerns us as human beings, interested in foreseeing the future with some degree of confidence on the basis of our scanty, imprecise and uncertain knowledge of the present and the past, all arguments about determinism are purely academic ... In the final analysis, it seems to be of very little consequence or assistance to us whether we take up a position for or against the plausibility of the hypothesis that Laplace's superman could work out the entire future *if only he knew the entire present in every detail.*

8   Probabilismo redux

De Finetti did not come to embrace indeterminism as a result of his study of the foundations of quantum physics. To the contrary. It was, instead, an integral part of his scientific philosophy from very early on. In "Probabilismo", de Finetti bluntly states in its first section (de Finetti 1931: 169): "we cannot accept determinism" because "it lacks all meaning". In the relativistic conception of science (to which de Finetti subscribed),

> Nature will not appear to it as a monstrous and incorrigibly exact clockwork mechanism where everything that happens is what must happen because it could not but happen, and where all is foreseeable if one knows how the mechanism works. (p. 170)

This rejection of determinism preceded quantum mechanics, but "in the struggle against determinism, modern physics is in fact our ally" (p. 216).

Why an ally? De Finetti imagines someone (our causal subjectivist introduced earlier) who believes in the existence of causal laws but regards probability as a tool for making subjective judgments absent the information needed to see outcomes with certainty. Modern physics helps combat such a position because it makes more uncertain and indefinite the position of those who might want, as I have supposed, to abandon the objective value of probability but not the objective value of causality. Between necessity and probability, between causality and randomness, where shall the boundary be drawn? (cf. von Mises 1957: 209-211)

Now we see the complete disconnect between de Finetti and subjectivists such as Good who regard the indeterminism of quantum physics as potentially introducing a novel element in science where physical probabilities become inescapable. For de Finetti modern physics topples a second idol – objective cause – rather than restoring the first idol of objective chance!

Of course perhaps the wheel will turn, and just as the statistical laws of today are replacing the deterministic laws of the past, perhaps in the physics of the future deterministic laws may once again reign supreme (read hidden variables). De Finetti replies that at the fundamental level of atoms and molecules, "the function of probabilistic concepts is deeper", and he then quotes with (apparent) approbation something he heard at a conference in 1929 (pp. 222-223, fn. 42):

> Radium atoms decay at successive times, with a half-life of about two thousand years. So, in a piece of radium there are atoms that will decay in the next minute, and others that will decay only in thousands of centuries. The new theoreticians claim that one cannot find, and indeed that there does not exist, any difference among these kinds of atoms, that have such different lifetimes, or any difference in the ambient conditions surrounding their decay. The decay of one or the other is purely random.

This passage makes it clear that de Finetti saw no contradiction between the "purely random" nature of radioactive decay, and the subjective nature of the probability of decay. How is this possible? The answer should by now be clear: in order for something to be "objective", it must be something one can observe or measure. Can one observe or measure the probability of radioactive decay? No: *single-case propensities are metaphysical constructs, and infinite limiting frequencies are fictions.* All one actually *observes* is a finite set of outcomes or measurements made over a finite period of time, providing us with statistical data that is used to arrive at previsions as we strive to cope with the pervasive uncertainties of life.

For further discussion of de Finetti's views on indeterminism and statistical regularities, see de Finetti (1931b), and its discussion in von Plato (1994: 258-264).

9 Conclusion

In order to understand de Finetti's views on quantum physics, one must start at the very beginning ("Probabilismo"), not the very end (*Teoria delle Probabilità*). De Finetti's subjectivism was a manifestation of his relativism, a perspective that informed not only de Finetti's general philosophy of science, but also his general philosophy of life ("Probabilismo", Section 32; see also Jeffrey 1989). Central to this viewpoint was de Finetti's absolute rejection of determinism in all its forms. Modern physics was merely an "ally" in this struggle, but not its motive force. The fact that the "new theoreticians" thought radioactive decay "purely random" merited a footnote in "Probabilismo", no more. For de Finetti the probabilities of quantum physics are ultimately no different than any other, no more objective than any other.

In his 1959 Varenna lectures de Finetti briefly touches on quantum physics, but a different aspect – the phenomenon of "complementarity", the existence of events that are not simultaneously verifiable. Like Koopman, de Finetti regarded the idea that this was somehow paradoxical as merely a confusion of thought, no more; such event pairs are common in everyday life. In such matters even the great von Mises could fall victim to "curious misunderstandings" (de Finetti 1972: 208).

But as Suppes notes, there *is* a very real difference between the prosaic and the quantum cases: one might – as de Finetti did – reject outright the existence of determinism in our everyday lives on purely philosophical grounds as merely a useless metaphysical anachronism, but quantum physics claimed to be able to actually *demonstrate* its impossibility. Surely this was interesting!

And so de Finetti began to read further in the literature of quantum physics. De Finetti found von Neumann's arguments about the impossibility of hidden variables convincing, but was intrigued at the same time by the novel structure of quantum logic. But while von Neumann saw the logical problems arising from complementarity as specific to quantum mechanics, another author – Bodiou – regarded them as much more pervasive, a viewpoint de Finetti obviously found more *simpatico*. It is this new and extended logic that de Finetti discusses at length in his Appendix.

The entire episode illustrates very clearly the "radical" nature of de Finetti's subjective probability. For I. J. Good, quantum physics raises the issue of whether physical probabilities – even if absent in the classical universe – might exist at the quantum level. For de Finetti – if his philosophical views are fully appreciated – the question does not even arise.

10 Coda: back to the future

> The speaker will meditate on what he likes and what he dislikes about taking a Bayesian view of quantum theoretic probabilities. The talk will be very informal if only because, as we all know, there is no quantum world. – David Mermin

Recently, subjective interpretations of quantum mechanics have generated increasing interest. Two manifestations of this interesting development are the conference, "Being Bayesian in a Quantum World", held at the University of Konstanz, August 1-5, 2005, and an issue of *Studies in History and Philosophy of Modern Physics* (2007, volume 38, number 2) devoted to "Probabilities in quantum mechanics". The first two papers in it, Bub (2007) and Caves et al. (2007), discuss two different Bayesian interpretations of quantum mechanics, and afford at least a partial entry into some of the relevant current literature.

ACKNOWLEDGEMENTS. I thank Maria Carla Galavotti for the invitation to speak at the 2006 de Finetti conference in Bologna, when a preliminary version of this paper was presented, and Eugenio Regazzini for some very perceptive and helpful comments in the discussion period afterwards.

References

Armendt, B. (1993). "Dutch Books, Additivity, and Utility Theory". *Philosophical Topics* 21, pp. 1-20.

Bodiou, G. (1964). *Théorie dialectique des probabilities*. Paris: Gauthier-Villars.

Bub, J. (2007). "Quantum Probabilities as Degrees of Belief". *Studies in History and Philosophy of Modern Physics* 38, pp. 232-254.

Bayes, T. (1764). "An Essay towards Solving a Problem in the Doctrine of Chances". *Philosophical Transactions of the Royal Society of London for 1763* 53, pp. 370-418.

Caves, C. M., Fuchs, C. A. and Schack, R. (2007). "Subjective Probability and Quantum Certainty". *Studies in History and Philosophy of Modern Physics* 38, pp. 255-274.

Cifarelli, D. M. and Regazzini, E. (1996). "De Finetti's Contribution to Probability and Statistics". *Statistical Science* 11, pp. 253-282.

de Finetti, B. (1931a). "Probabilismo. Saggio critico sulla teoria della probabilità e sul valore della scienza". *Logos* 14, pp. 163-219. (English translation, "Probabilism", *Erkenntnis* 31 (1989), pp. 169-223; page references are to this edition.)

de Finetti, B. (1931b). "Le leggi differenziali e la rinunzia al determinismo". *Rendiconti del Seminario Matematico della R. Università di Roma* 7, pp. 63-74. (English translation, "Differential Laws and the Renunciation of Determinism", in *Induction and Probability*, ed. by P. Monari and D. Cocchi, Bologna: CLUEB, 1993, pp. 323-334.)

de Finetti, B. (1937). "La prévision: ses lois logiques, ses sources subjectives". *Annales de l'Institut Henri Poincaré* 7, pp. 1-68. (English translation, "Foresight. Its logical laws, Its subjective sources", in *Studies in Subjective Probability*, ed. by H. E. Kyburg, Jr. and H. E. Smokler, New York: John Wiley & Sons, 1964, pp. 93-158.)

de Finetti, B. (1970). *Teoria delle Probabilità*. Turin: Einaudi. (English translation, de Finetti, 1974 and 1975.)

de Finetti, B. (1972). *Probability, Induction, and Statistics*. New York: Wiley.

de Finetti, B. (1974). *Theory of Probability*, vol. 1. New York: Wiley.

de Finetti, B. (1975). *Theory of Probability*, vol. 2. New York: Wiley.

de Finetti, B. (1979). "Probability and Exchangeability from a Subjective Point of View". *International Statistical Review / Revue internationale de Statistique* 47, pp. 129-135.

Dodds, E. R. (1977). *Missing Persons*. Oxford: Clarendon Press.

Fienberg, S. E. (1971). "Randomization and Social Affairs: The 1970 Draft Lottery". *Science* 171, pp. 255-261.

Good, I. J. (1965). *The Estimation of Probabilities*. Cambridge, MA: MIT Press.

Good, I. J. (1977). "Review of de Finetti, 1974 and 1975". *Bulletin of the American Mathematical Society* 83, pp. 94-97.

Hacking, I. (1965). *Logic and Statistical Inference*. Cambridge: Cambridge University Press.

Hacking, I. (1971). "Equipossibility Theories of Probability". *British Journal for the Philosophy of Science* 22, 339-355.

Holland, P. W. (1986). "Statistics and Causal Inference". *Journal of the American Statistical Association* 81, pp. 945-960.

Jeffrey, R. C. (1977). "Mises Redux". *Basic Problems in Methodology and Linguistics*, ed. by R. Butts and J. Hintikka, Dordrecht: Reidel, 1977, pp. 213-222.

Jeffrey, R. C. (1986). "Judgmental Probability and Objective Chance". *Erkenntnis* 24, pp. 5-16.

Jeffrey, R. C. (1989). "Reading 'Probabilismo'". *Erkenntnis* 31, pp. 225-237.

Kemp, A. W. and Kemp, C. D. (1991). "Weldon's Dice Data Revisited". *The American Statistician* 45, pp. 216-222.

Koopman, B. (1957). "Quantum Theory and the Foundations of Probability". In *Proceedings of Symposia in Applied Mathematics, Volume VII: Applied Mathematics*, ed. by L. A. MacColl, New York: McGraw-Hill, pp. 97-102.

Laplace, P.-S. (1814). *Essai philosophique sur les probabilités*. Paris: Courcier. (English translation, *Philosophical Essay on Probabilities*, trans. by A. I. Dale, New York: Springer-Verlag, 1995. Page reference to this edition.)

Lewis, D. (1980). "A Subjectivist's Guide to Objective Chance". In *Studies in Inductive Logic and Probability*, vol. II, ed. by R. C. Jeffrey, Berkeley: University of California Press, pp. 262-292.

Mises, R. von (1957). *Probability, Statistics, and Truth*. London: George Allen & Unwin, Ltd.

Ramsey, F. P. (1931). *The Foundations of Mathematics and Other Logical Essays*, ed. by R. B. Braithwaite. London: Routledge and Kegan Paul.

Reichenbach, H. (1944). *Philosophical Foundations of Quantum Mechanics*. Berkeley: University of California Press.

Suppes, P. (1961). "Probability Concepts in Quantum Mechanics". *Philosophy of Science* 28, 378-389.

Suppes, P. and Zanotti, M. (1989). "Conditions on Upper and Lower Probabilities to Imply Probabilities". *Erkenntnis* 31, pp. 323-345.

von Neumann, J. (1955). *Mathematical Foundations of Quantum Mechanics*. Princeton: Princeton University Press.

von Neumann, J. and Birkhoff, G. (1936). "The Logic of Quantum Mechanics". *Annals of Mathematics* 37, pp. 823-843.

von Plato, J. (1994). *Creating Modern Probability*. Cambridge-New York: Cambridge University Press.

Zabell, S. L. (1991). "Ramsey, Truth, and Probability". *Theoria* 57, pp. 211-238.

Zabell, S. L. (1997). "The Continuum of Inductive Methods Revisited". In *The Cosmos of Science: Essays of Exploration*, ed. by J. Earman and J. D. Norton, University of Pittsburgh Press/Universitäts Verlag-Konstanz, pp. 351-38.

5

Diachronic Coherence
and Radical Probabilism[*]

BRIAN SKYRMS

1   Introduction

In 1937, Bruno de Finetti founded the theory of personal (subjective) probability on the concept of *coherence* – immunity from a Dutch book. The concept of coherence was that of coherence of degrees of belief at a single time, *synchronic coherence.* It is perhaps arguable that there is a theory of coherent belief change implicit in de Finetti's treatment of conditional probability. But it was still possible for Hacking to argue 30 years later that subjective Bayesians had no explicit theory of coherence across time, on which to found belief change by conditioning on the evidence. Explicit discussion of coherence across time, *diachronic coherence*, was inaugurated in the philosophical community by David Lewis' Dutch Book argument for conditioning[1], and in the statistical community by Freedman and Purves' (1969) argument for updating using Bayes' theorem.

Richard Jeffrey advocated a flexible theory of personal probability that is open to all sorts of learning situations. He opposed what he saw as the use of a conditioning model as an epistemological straitjacket in the work of Clarence Irving Lewis (1946). Lewis's dictum "No probability without certainty" followed from the idea that probabilities must be up-

[*] This is a slightly modified version of the article appeared in *Philosophy of Science* 73, 2006, pp. 959-968. Published here with permission of the Philosophy of Science Association.

[1] Reported in Teller (1973).

Bruno de Finetti, Radical Probabilist
Maria Carla Galavotti (ed.)
Copyright © 2009

dated by conditioning on the evidence. Jeffrey's *probability kinematics* – now also known as "Jeffrey conditioning" – provided an alternative (Jeffrey 1957, 1965, 1968).

It was not meant to be the only alternative. Jeffrey articulated a philosophy of *radical probabilism* that held the door open to modeling all sorts of epistemological situations. In this spirit, I will look at diachronic coherence from a point of view that embodies minimal epistemological assumptions and then add constraints little by little.

2 Arbitrage

There is a close connection between Bayesian coherence arguments and the theory of arbitrage (Shin 1992). Suppose we have a market in which a finite number[2] of assets are bought and sold. Assets can be anything: stocks and bonds, pigs and chickens, apples and oranges. The market determines a unit price for each asset, and this information is encoded in a price vector $\mathbf{x} = <x_1, ..., x_n>$. You may trade these assets today in any (finite) quantity. You are allowed to take a short position in an asset; that is to say, you sell it today for delivery tomorrow. Tomorrow, the assets may have different prices, $\mathbf{y}_1, ..., \mathbf{y}_m$. To keep things simple, we initially suppose that there are a finite number of possibilities for tomorrow's price vector. A *portfolio*, $\mathbf{p}$, is a vector of real numbers that specifies the amount of each asset you hold. Negative numbers correspond to short positions. You would like to *arbitrage the market*, that is, to construct a portfolio today whose cost is negative (you can take out money) and such that tomorrow its value is nonnegative (you are left with no net loss), no matter which of the possible price vectors is realized.

According to the *fundamental theorem of asset pricing*, you can arbitrage the market if and only if the price vector today falls outside the convex cone spanned by the possible price vectors tomorrow.[3]

There is a short proof that is geometrically transparent. The value of a portfolio, $\mathbf{p}$, according to a price vector, $\mathbf{y}$, is the sum over the assets of quantity times price: the dot product of the two vectors. If the vectors are

---

[2] We keep things finite at this point because we want to focus on diachronic coherence and avoid the issues associated with the philosophy of the integral.

[3] If we were to allow an infinite number of states tomorrow, we would have to substitute the closed convex cone generated by the possible future price vectors.

orthogonal, the value is zero. If they make an acute angle, the value is positive; if they make an obtuse angle, the value is negative. An arbitrage portfolio, $\mathbf{p}$, is one such that $\mathbf{p} \cdot \mathbf{x}$ is negative and $\mathbf{p} \cdot \mathbf{y}_i$ is nonnegative for each possible $\mathbf{y}_i$; $\mathbf{p}$ makes an obtuse angle with today's price vector and is orthogonal or makes an acute angle with each of the possible price vectors tomorrow. If $\mathbf{p}$ is outside the convex cone spanned by the $\mathbf{y}_i$'s, then there is a hyperplane that separates $\mathbf{p}$ from that cone. An arbitrage portfolio can be found as a vector normal to the hyperplane. It has zero value according to a price vector on the hyperplane, a negative value according to today's prices, and a nonnegative value according to each possible price tomorrow. On the other hand, if today's price vector is in the convex cone spanned by tomorrow's possible price vectors, then (by Farkas's lemma) no arbitrage portfolio is possible.

Suppose, for example, that the market deals in only two goods, apples and oranges. One possible price vector tomorrow is $1 for an apple, $1 for an orange. Another is that an apple will cost $2, while an orange is $1. These two possibilities generate a convex cone, as shown in Figure 1a. (We could add lots of intermediate possibilities, but that wouldn't make any difference to what follows.)

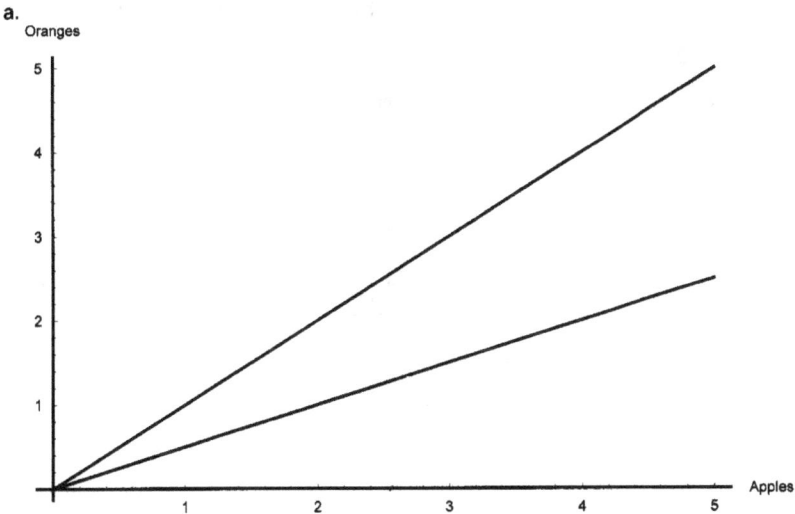

Figure 1a.

Let's suppose that today's price vector lies outside the convex cone, say apples at $1, oranges at $3. Then it can be separated from the cone by a hyperplane (in two dimensions, a line), for example, the line oranges p 2 apples, as shown in Figure 1b.

b.

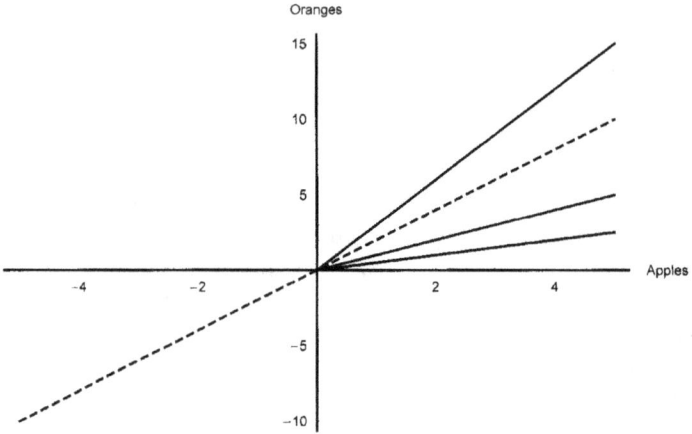

Figure 1b.

Normal to that hyperplane we find the vector <2 apples, -1 orange >, as in Figure 1c.

c.

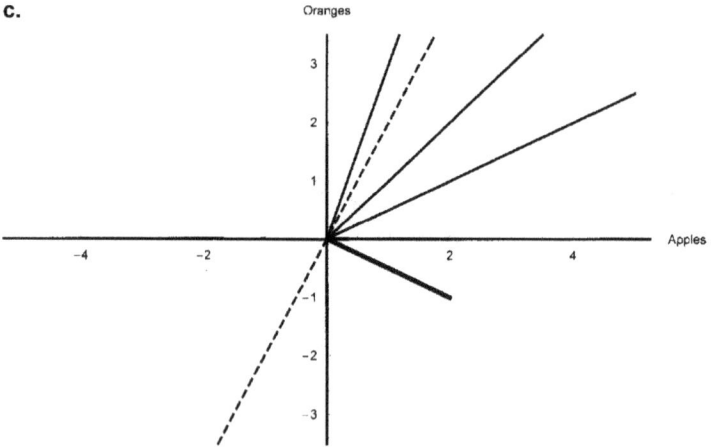

Figure 1c.

This should be an arbitrage portfolio, so we sell one orange short and use the proceeds to buy two apples. But at today's prices, an orange is worth

$3; so we can pocket a dollar, or – if you prefer – buy three apples and eat one.

Tomorrow we have to deliver an orange. If tomorrow's prices were to fall exactly on the hyperplane, we would be covered. We could sell our two apples and use the proceeds to buy the orange. But in our example, things are even better. The worst that can happen tomorrow is that apples and oranges trade one-to-one, so we might as well eat another apple and use the remaining one to cover our obligation for an orange. The whole business is straightforward: sell dear, buy cheap. Notice that at this point there is no probability at all in the picture.

3   Degrees of belief

In the foregoing, assets could be anything. As a special case they could be tickets paying $1 if p, nothing otherwise, for various propositions, p. The price of such a ticket can be thought of as the market's collective *degree of belief* or *subjective probability* for p. We have not said anything about the market except that it will trade arbitrary quantities at the market price. The market might or might not be implemented by a single individual – the bookie of the familiar Bayesian metaphor.

Without yet any commitment to the nature of the propositions involved, the mathematical structure of degrees of belief, or the characteristics of belief revision, we can say that arbitrage-free degrees of belief today must fall within the convex cone spanned by the degrees of belief tomorrow. This is the fundamental diachronic coherence requirement. Convexity is the key to everything that follows.

4   Probability

Suppose, in addition, that the propositions involved are true or false and that tomorrow we learn the truth. (We also assume that we can neglect discounting the future. A guarantee of getting $1 tomorrow is as good as getting $1 today.) Then tomorrow a ticket worth $1 if p, nothing otherwise, would be worth either $1 or $0 depending on whether we learn whether p is true or not.

And suppose that we have three assets being traded that have a logical structure. There are tickets worth $1 if p, nothing otherwise; $1 if q, nothing otherwise; and $1 if p or q, nothing otherwise. Furthermore, p and q are incompatible. This additional structure constrains the possible price vectors tomorrow, so that the convex cone becomes the two-dimensional object $z = x + y$ (x, y nonnegative), as shown in Figure 2:

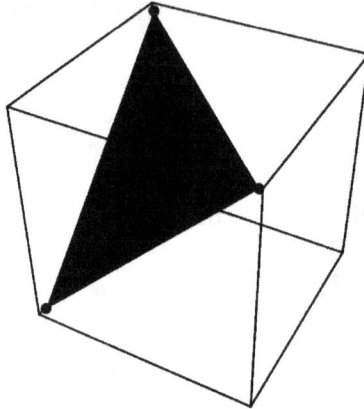

Figure 2.

Arbitrage-free degrees of belief must be additive. *Additivity of subjective probability* comes from the *additivity of truth value* and the fact that *additivity is preserved under convex combination*. One can then complete the coherence argument for probability by noting that coherence requires a ticket that pays $1 if a tautology is true to have the value $1.

Notice that from this point of view, the synchronic Dutch books are really special cases of diachronic arguments. You need the moment of truth for the synchronic argument to be complete. The assumption that there is such a time is a much stronger assumption than anything that preceded it in this development.

An intuitionist, for example, may have a conception of proposition and of the development of knowledge that does not guarantee the existence of such a time, even in principle. Within such a framework, coherent degrees of belief need not obey the classical laws of the probability calculus.

5 Probabilities of probabilities

Today the market trades tickets that pay $1 if p_i, nothing otherwise, where the p_i's are some "first-order" propositions. All sorts of news comes in, and tomorrow the price vector may realize a number of different possibilities. (We have not, at this point, imposed any model of belief change.) The price vector for these tickets tomorrow is itself a fact about the world, and there is no reason why we could not trade in tickets that pay off $1 if tomorrow's price vector is **p** or if tomorrow's price vector is in some set of possible price vectors, for the original set of propositions. The prices of these tickets represent subjective probabilities today about subjective probabilities tomorrow.

Some philosophers have been suspicious about such entities, but they arise quite naturally. And in fact, they may be less problematic than the first-order probabilities over which they are defined. The first-order propositions, p_i, could be such that their truth value might or might not ever be settled. But the question of tomorrow's price vector for unit wagers over them is settled tomorrow. Coherent probabilities of tomorrow's probabilities should be additive, no matter what.

6 Diachronic coherence revisited

Let us restrict ourselves to the case in which we eventually do find out the truth about everything and all bets are settled (perhaps on Judgment Day), so degrees of belief today and tomorrow are genuine probabilities. We can now consider tickets that are worth $1 if the probability tomorrow of $p = a$ and p, nothing otherwise, as well as tickets that are worth $1 if the probability tomorrow of $p = a$.

These tickets are logically related. Projecting to the two dimensions that represent these tickets, we find that there are only two possible price vectors tomorrow. Either the probability tomorrow of p is not equal to a, in which case both tickets are worth nothing tomorrow, or the probability tomorrow of p is equal to a, in which case the former ticket has a price of a and the latter has a price of $1. The cone spanned by these two vectors is just a ray as shown in Figure 3.

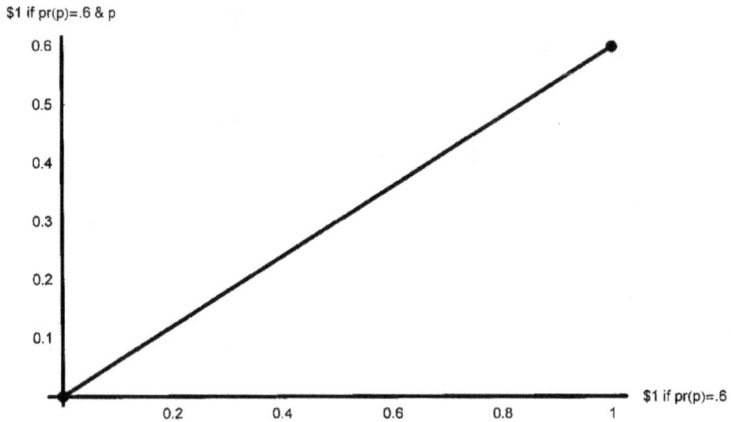

$1 if pr(p)=.6 & p

Figure 3.

So today, the ratio of these two probabilities (provided that they are well defined) is a. In other words, today the conditional probability of p, given that the probability tomorrow of $p = a$, is a. It then follows that to avoid a Dutch book, the probability today must be the expectation of the probability tomorrow (Goldstein 1983; van Fraassen 1984). Since convexity came on stage, it has been apparent that this expectation principle has been waiting in the wings. The introduction of probabilities of probabilities allows it to be made explicit.

7   Coherence and conditioning

In accord with Jeffrey's philosophy of radical probabilism, we have imposed no restrictive model of belief change. A conditioning situation is allowed, but not required. That is to say, there may be first-order propositions, $e_1, ..., e_n$, that map one-to one to possible degrees of belief tomorrow, $q_1, ..., q_n$, such that, for our degrees of belief today, $\mathbf{p}$, and for all propositions under consideration, s, $\mathbf{q}i\ (s) = \mathbf{p}(s$ given $e_i)$, in which case we have a conditioning model. But there need not be such propositions, which is the case that radical probabilism urges us not to ignore. In this case, convexity still provides an applicable test of diachronic coherence.

On the other hand, with the introduction of second-order probabilities, coherence *requires* belief change by conditioning, that is to say, conditioning on propositions about what probabilities will be tomorrow (see Skyrms 1980; Good 1981). These are, of course, quite different from the first-order sense-data propositions that C. I. Lewis had in mind.

8   Probability kinematics

Where does Richard Jeffrey's probability kinematics fit into this picture? Belief change by kinematics on some partition is not sufficient for diachronic coherence. The possible probability vectors tomorrow may have the same probabilities conditional on p and on its negation as today's probability without today's probability of p being the expectation of tomorrow's. Diachronic coherence constrains probability kinematics.

In a finite setting, belief change is always by probability kinematics on *some* partition, the partition whose members are the atoms of the space. But, as Jeffrey always emphasized, coherent belief change need not consist of probability kinematics on some nontrivial partition. That conclusion follows only from stronger assumptions that relate the partition in question to the learning situation.

Suppose that between today and tomorrow we have a learning experience that changes the probability of p, but not to zero or one. And suppose that then, by the day after tomorrow, we learn the truth about p. We can express the *assumption* that we have gotten information only *about p* on the way through tomorrow to the day after tomorrow by saying that we move from now to then by conditioning on p or on its negation. This is the assumption of *sufficiency* of the partition $\{p, \text{-}p\}$. Then one possible probability tomorrow has the probability of p as one and the probability of p & q as equal to today's Pr (q given p) and the other possible probability tomorrow has the probability of p as zero and the probability of $\text{-}p$ & q as equal to today's Pr (q given $\text{-}p$). This is shown in Figure 4 for Pr (q given p) = .9 and Pr(q given $\text{-}p$) = .2.

Diachronic coherence requires that tomorrow's probabilities must fall on the line connecting the two points representing possible probabilities the day after tomorrow and thus must come from today's probabilities by kinematics on $\{p, \text{-}p\}$. This is the basic diachronic coherence argument that in my earlier work (Skyrms 1987, 1990) was cloaked in concerns about infinite spaces.

As Jeffrey always emphasized, without the assumption of sufficiency of the partition, there is no coherence argument. But equally, if there is no assumption that we learn just the truth of p, there is no argument for conditioning on p in the case of certain evidence.

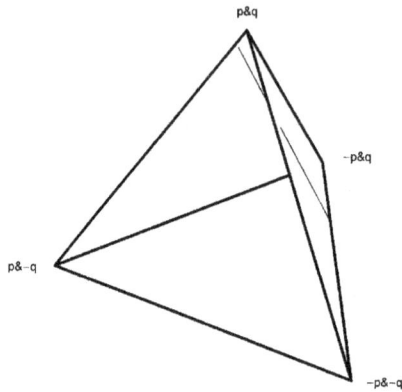

Figure 4.

9 Tomorrow and tomorrow and tomorrow

Consider not two or three days, but an infinite succession of days. Assume that degrees of belief are all probabilities. The probability of p tomorrow, the probability of p the day after tomorrow, the probability of p the next day, and so forth are a sequence of random variables. Diachronic coherence requires that they form a *martingale* (see Skyrms 1996; Zabell 2002). Richer cases lead to vector-valued martingales.

10 Diachronic coherence generalized

Looking beyond the scope of this paper, suppose that you throw a point dart at a unit interval, and the market can trade in tickets that pay $1 if it falls in a certain subset, $0 otherwise. This is something of an idealization to say the least, and the question arises as to how coherence might be applied. One natural idea might be to idealize the betting situation so as to allow a countable number of bets, in which case coherence requires countable additivity, a restriction of contracts to measurable sets, and, in general, the orthodox approach to probability. Then the martingale convergence theorem applies: Coherence entails convergence.

An approach more faithful to the philosophy of de Finetti would allow a finite number of bets at each time. This leads to *strategic measure*, a notion weaker than countable additivity but stronger than simple finite additivity (Lane and Sudderth 1984, 1985). Orthodox martingale theory uses countable additivity, but there is a finitely additive martingale theory built on strategic measures (Purves and Sudderth 1976). A version of the "coherence entails convergence" result can be recovered, even on this more conservative approach (Zabell 2002).

References

de Finetti, B. (1937). "La prévision: ses lois logiques, ses sources subjectives". *Annales de l'institut Henri Poincaré* 7, pp. 1 -68.

Freedman, D. and Purves, R. (1969). "Bayes' Method for Bookies". *The Annals of Mathematical Statistics* 40, pp. 1177-1186.

Goldstein, M. (1983). "The Prevision of a Prevision". *Journal of the American Statistical Association* 78, pp. 817-819.

Good, I. J. (1981). "The Weight of Evidence Provided from an Uncertain Testimony or from an Uncertain Event". *Journal of Statistical Computation and Simulation* 13, pp. 56-60.

Hacking, I. (1967). "Slightly More Realistic Personal Probability". *Philosophy of Science* 34, pp. 311-325.

Jeffrey, R. (1957). *Contributions to the Theory of Inductive Probability*. PhD dissertation, Princeton University.

Jeffrey, R. (1965). *The Logic of Decision*. New York: McGraw-Hill; 3rd rev. ed. Chicago: University of Chicago Press, 1983.

Jeffrey, R. (1968). "Probable Knowledge". In *The Problem of Inductive Logic*, ed. by I. Lakatos. Amsterdam: North-Holland.

Lane, D. A. and Sudderth, W. (1984). "Coherent Predictive Inference". *Sankhya* 46, pp. 166-185.

Lane, D. A. and Sudderth, W. (1985). "Coherent Predictions Are Strategic". *Annals of Statistics* 13, pp. 1244-1248.

Lewis, C. I. (1946). *An Analysis of Knowledge and Valuation*. LaSalle, IL: Open Court.

Purves, R. and Sudderth, W. (1976). "Some Finitely Additive Probability". *Annals of Probability* 4, pp. 259-276.

Shin, H. S. (1992). "Review of *The Dynamics of Rational Deliberation*". *Economics and Philosophy* 8, pp. 176-183.

Skyrms, B. (1980). "Higher Order Degrees of Belief". In *Prospects for Pragmatism*, ed. by D. H. Mellor. Cambridge: Cambridge University Press, pp. 109-138.

Skyrms, B. (1987). "Dynamic Coherence and Probability Kinematics". *Philosophy of Science* 54, pp. 1-20.

Skyrms, B. (1990). *The Dynamics of Rational Deliberation*. Cambridge, MA: Harvard University Press.

Skyrms, B. (1996). "The Structure of Radical Probabilism". *Erkenntnis* 45, pp. 285-297.

Teller, P. (1973). "Conditionalization, Observation and Change of Preference". *Synthese* 26, pp. 218-258.

van Fraassen, B. (1984). "Belief and the Will". *Journal of Philosophy* 81, pp. 235-256.

Zabell, S. (2002). "It All Adds Up: The Dynamic Coherence of Radical Probabilism". *Philosophy of Science* 69 (Proceedings), pp. S98-S103.

6

De Finetti's Subjectivism, Objective Probability, and the Empirical Validation of Probability Assessments

A. PHILIP DAWID AND MARIA CARLA GALAVOTTI

Foreword

A well known claim of Bruno de Finetti is that "Probability does not exist". This sentence appears in capital letters in the Preface to the English edition of his *Theory of Probability*[1], and reappears at the beginning of the article "Probabilità" in the *Enciclopedia Einaudi*.[2] Such a strong opposition to "objective" probability is obviously inspired by the desire to avoid metaphysical "contaminations" and is rooted in de Finetti's anti-realist and pragmatist philosophy of probability[3].

A bad consequence of de Finetti's statement is that of fostering the feeling that subjectivism is surrounded by a halo of arbitrariness. The suspicion that subjectivism represents an "anything goes" approach is actually shared by researchers from a variety of fields. In the first place, many physicists object to de Finetti's refusal to retain a notion of "objective" probability to be applied to science, and prefer to appeal to the frequency interpretation of probability. A similar attitude is embraced by, among others, forensic scientists, who are mostly suspicious of subjective probability[4] and often turn to logical probability, presumably reassured

---

[1] See de Finetti (1970a), English edition (1975).
[2] See de Finetti (1980).
[3] For a detailed discussion of de Finetti's anti-realist philosophy of probability, see Galavotti (1989).
[4] See for instance Laudan (2006).

Bruno de Finetti, Radical Probabilist
Maria Carla Galavotti (ed.)
Copyright © 2009

by the promise of objectivity conveyed by the term "logical".[5] Against this tendency we will argue that de Finetti's claim should *not* be taken to suggest that subjectivism is an anarchist approach according to which probability can take whatever value you like. De Finetti struggles against *objectivism*, or the idea that probability depends entirely on some aspects of reality, not against *objectivity*. He strongly opposes the "distortion" of "identifying objectivity and objectivism", deemed a "dangerous mirage" (de Finetti 1962b: 344), but does not deny that there is a *problem of objectivity* of evaluations of probability. In fact the problem was tackled by de Finetti in his usual insightful way, partly in collaboration with Leonard Jimmie Savage.[6]

The following pages are divided in two parts: the first surveys de Finetti's position regarding "objective" probability and the problem of objectivity, while the second focuses on tools for evaluating probability assessments.

Part I

In this part we focus on de Finetti's attitude towards subjectivism, objective probability and the issue of objectivity.

I.1 "Objective" probability

A first aspect of de Finetti's refusal to attach an "objective" meaning to probability that should be pointed out is that it impinges upon notions that are commonly used by scientists, such as those of *chance* and *physical probability*. No doubt, the lack of consideration for such notions is a gap in de Finetti's perspective and is responsible for the distrust that subjectivism inspires to a number of scientists. Spurred by his anti-realism, de Finetti never paid much attention to the use made of probability in science, convinced that science can be seen as a mere continuation of everyday life and subjective probability is all that is needed. Only the posthumous volume *Filosofia della Probabilità* (1995), containing the text, transcribed from tapes, of a course given by de Finetti in 1979, includes a few remarks that are relevant to the point. There de Finetti admits that probability distributions belonging to scientific theories – he refers specifically to statistical mechanics – can be taken as "more solid grounds"

---

[5] See for instance Stella (2003).
[6] See Savage (1971), where the collaboration between Savage and de Finetti on the subject is mentioned.

for subjective opinions (de Finetti 1995: 117). This allows for the conjecture that late in his life he entertained the idea that probabilities encountered in science derive a peculiar "robustness" from scientific theories.[7]

A similar attitude is to be found in the writings of Frank Ramsey, the other father of subjective probability, and Harold Jeffreys, the geophysicist and probabilist who embraced the notion of logical probability, but shared various aspects of subjectivism.[8] Both of these authors are convinced that notions like "chance" and "probability in physics" can be developed within an epistemic conception of probability. For Ramsey this is done by referring such notions to *systems of beliefs* that typically include widely accepted generalisations such as laws of nature. More specifically, probabilities occurring in physics, which are actually derived from scientific theories, are regarded as belonging to systems of belief that include such theories. They can be seen as "ultimate chances", in the sense that within the theoretical framework in which they occur there is no way of replacing them with deterministic laws. Their objective character derives from the objectivity ascribed to theories, which is warranted by acceptance on the part of the scientific community: only those theories which gain *universal assent* in the long run are accepted and taken as true. In conclusion, for Ramsey chance attributions indicate a way in which beliefs in various facts belonging to science are guided by scientific theories.[9] Although de Finetti did not quite go that far, the remarks reported in *Filosofia della Probabilità* seem to point to a similar direction.

I.2   The problem of objectivity

As already observed, de Finetti addresses the issue of objectivity, rephrased in subjectivistic terms as the problem of good probability appraisal. In order to understand his position one should keep in mind the distinction between the *definition* and the *evaluation* of probability. These are seen by de Finetti as utterly different concepts which should not be conflated. In fact he believes that the confusion between the definition and the evaluation of probability imprints all the other interpretations of probability, namely frequentism, logicism and the classical approach. Upholders of these viewpoints look for a unique criterion – be it frequency, or symmetry – and use it as grounds for both the definition

---

[7] This is argued in some detail in Galavotti (2001) and (2005).

[8] See Galavotti (2003) for an overview of Jeffreys' epistemology.

[9] For more details on Ramsey's view of chance and probability in physics see Galavotti (1995) and (1999).

and the evaluation of probability. In so doing, they embrace a "rigid" attitude towards probability, which consists "in defining (in whatever way, according to whatever conception) the probability of an event, and in univocally determining a function" (de Finetti 1933: 740).

By contrast, subjectivism does not conflate the definition of probability with its evaluation. Probability is *defined* as the *degree of belief* "as actually held by someone, on the ground of his whole knowledge, experience, information" (de Finetti 1968: 45) regarding an event whose outcome is uncertain. For subjectivists there are no "correct" probability assignments; in de Finetti's words:

> The subjective theory ... does not contend that the opinions about probability are uniquely determined and justifiable. Probability does not correspond to a self-proclaimed "rational" belief, but to the effective personal belief of anyone. (de Finetti 1951: 218)

In view of this, the subjective viewpoint is deemed "elastic": instead of committing the choice of one particular function to a single rule or method, de Finetti regards it as the result of a complex and largely context-dependent procedure, which necessarily involves subjective elements. In other words, the *evaluation* of probability should take into account all available evidence, including frequencies and symmetries. However, it would be a mistake to put these elements, which are useful ingredients of the evaluation of probability, at the basis of the definition of probability.

De Finetti maintains that

> every probability evaluation essentially depends on two components: (1) the objective component, consisting of the evidence of known data and facts; and (2) the subjective component, consisting of the opinion concerning unknown facts based on known evidence. (de Finetti 1974: 7)

In other words, objective elements do provide a basis for judgment, though *not* the *only* basis. The explicit recognition of the role played by subjective elements within the formation of probability judgments is for de Finetti a prerequisite for the appraisal of objective elements. As he puts it:

> Subjective elements will in no way destroy the objective elements nor put them aside, but bring forth the implications that originate

only after the conjunction of both objective and subjective elements at our disposal. (de Finetti 1973: 366)

Attention is called to the fact that the collection and exploitation of factual evidence, which is the objective component of probability judgments, is intrinsically context-dependent and involves subjective elements of various kinds. Evidence must be collected carefully and skillfully, its exploitation depending on the judgment as to what elements are relevant to the problem under consideration and enter into the evaluation of probabilities. In addition, the collection and exploitation of evidence depends on economic considerations that vary according to the context. In practical situations a number of other factors influence probability evaluations, including the degree of competence of the evaluator, his optimistic or pessimistic attitudes, the influence exercised by most recent facts, and the like.[10] Therefore the evaluation of probability can only be seen as a highly complex procedure, resulting from the concurrence of all sorts of elements.

Equally subjective for de Finetti is the decision on how to let belief be influenced by objective elements. Typically, one relies on information regarding frequencies. As already pointed out, frequencies, like symmetry considerations, are useful and important ingredients of probability evaluations, *provided* that they are not used uncritically as automatic rules and conflated with probability. Keeping in mind the distinction between the definition and the evaluation of probability, one can make good use of frequencies to guide probability evaluations. Within de Finetti's perspective, the interaction between degrees of belief and frequencies rests on exchangeability. Assuming subjective exchangeability, whenever a considerable amount of information on frequencies is available this will strongly constrain probability assignments. But information on frequencies is often scant, and in this case the problem of how to obtain good probability evaluations becomes crucial.

In conclusion, once it has been acknowledged that probability is subjective and that there is no unique "rational" way of assessing probability, room can be made for a whole array of elements to influence and improve probability evaluations. This problem was addressed by de Finetti from the Sixties onwards. The approach adopted is based on scoring rules, such as "Brier's rule", named after the meteorologist who applied it to weather forecasts (Brier 1950). The nature of this method will be de-

---

[10] See de Finetti (1970b: 141-142).

scribed in the second part of the present paper, where it will be pointed out that it serves three purposes. In the first place it provides a tool for defining (subjective) probability in an operational fashion. As a matter of fact, in his late works de Finetti preferred the definition in terms of scoring rules to the "classical" one based on the betting scheme.[11] In addition, the method of scoring also allows to measure uncertainty, and can suggest a proper "punishment for bad behaviour". De Finetti calls attention to the fact that scoring rules can be used for improving probability evaluations made both by a single person and by several people, namely they provide a tool for enhancing "self-control" as well as a "comparative control" (de Finetti 1980: 1151). In order to test the goodness of this method de Finetti conducted an experiment among his students, who were asked to make forecasts on the results of soccer matches in the Italian championship. Unfortunately he did not publish the results of this experiment.[12]

As pointed out by de Finetti the method can be given a straightforward interpretation in tune with subjectivism. In his words:

> The objectivists, who reject the notion of personal probability because of the lack of verifiable consequences of any evaluation of it, are faced with the question of admitting the value of such a "measure of success" as an element sufficient to soften their forejudgments. The subjectivists, who maintain that a probability evaluation, being a measure of someone's beliefs, is not susceptible of being proved or disproved by the facts, are faced with the problem of accepting some significance of the same "measure of success" as a measure of the "goodness of the evaluation". (de Finetti 1962a: 360)

The following remark further clarifies this point:

> though maintaining the subjectivist idea that no fact can prove or disprove belief – de Finetti writes – I find no difficulty in admitting that any form of comparison between probability evaluations (of myself, of other people) and actual events may be an element influencing my further judgment, of the same status as any other kind of information (*ibidem*).

---

[11] See for instance de Finetti (1995).

[12] The experiment is mentioned in de Finetti (1962a). More material is to be found among the archival documents of the de Finetti Collection belonging to the Archives of Scientific Philosophy at the Hillman Library of the University of Pittsburgh.

To sum up, even though the intrinsic context-sensitivity of probability evaluation makes the idea of *absolute objectivity* nonsensical, a problem of objectivity arises, and is addressed by de Finetti.

Part II

In this part we describe an objective analysis of probability fully concordant with de Finetti's anti-realist philosophy. For further detail see Dawid (1986).

II.1   A proper scoring rule

Suppose a forecaster F is required to quote a probability value, q, for some uncertain event A. After the outcome is observed, F receives a *penalty score*: $(1-q)^2$ if A occurs, q^2 if not. Using notation favoured by de Finetti[13], we can also consider A as a binary variable, taking value 1 if the event occurs, 0 if not. Then the penalty can be written as $(A-q)^2$. It represents the (negative) pay-off in a game between F and Nature, with F selecting q, and then Nature selecting A. The game can also be described by means of its *pay-off table*, as in Table 1 (where the values have been multiplied by 100). This particular penalty function is called the *Brier*, or *quadratic*, scoring rule (Brier 1950).

Suppose F believes that the probability of A is p: how should she play the game? General principles of rational behaviour suggest that she should choose q to minimise her expected penalty score.[14] It is readily calculated that the expected penalty for choice q, when F truly believes the probability of A is p, is $S(p,q) = p(1-p) + (p-q)^2$. And this is uniquely minimised for $q=p$: that is to say, honesty is the best policy. This remarkable property is expressed by saying that the Brier score is *strictly proper*.

There are many other ways of defining a strictly proper scoring rule: for example, the *logarithmic scoring rule* (Good 1952), with pay-off table as displayed in Table 2. Savage (1971) showed that it would be incoherent to assign different probabilities when different scoring rules are used. For simplicity we shall here consider only the Brier score, but (with appropriate adjustments) virtually all the discussion and analysis below

---

[13] See de Finetti (1970a), Volume 1, Section 2.3.4.
[14] See Savage (1954).

would remain relevant if we were to base it on some other strictly proper scoring rule.

II.2   Three roles for penalty functions

There are three distinct purposes for which the Brier scoring rule can be used:

1) As a mathematical tool to define and analyse the *coherence* of a collection of numerical probability assessments (marginal and/or conditional) for various uncertain events. If there exists another such collection that would lead to a smaller score no matter which events occur, the original collection of assessments is said to be *incoherent*; else *coherent*. Using this definition de Finetti (1970a) demonstrates that any coherent specification must obey the standard rules (convexity, addition, product) of probability theory. But, as stressed in Part I, this approach gives no reason to prefer any one coherent specification to any other.

2) As a measuring instrument for quantifying uncertainty. A forecaster F faced with the task of assessing her probability for some event A might be presented with the pay-off table of Table 1. Her task is to choose one of the rows, which then commits her to receive the penalty in column 1 should A occur, or that in column 2 should it not. If she is pretty sure of A she will prefer a lower row, if pretty sure of its negation an upper row, and if unsure some row in the middle. Once her choice is made, the label attached to that row is her numerical probability for A – because of the strict propriety of the Brier score.

3) As a "punishment for bad behaviour". After F has announced her probability p for A (so determining a row of Table 1), Nature determines whether or not A happens (so determining a column). Finally F is actually penalised, according to the value at the intersection of that row and column.

Usage (1), although fundamental to de Finetti's theoretical development, will not be our concern here. As for usages (2) and (3), these can be considered as opposite sides of the same coin: (2) looking forwards, and (3) looking back. This is very similar to the dual role of penalties in the judicial system. On the one hand, the very existence of the tariff of penalties is intended to concentrate the mind, and thus encourage individuals to refrain from criminal activity (for this to work it is essential that the individuals in question be sentient and rational): this is similar to usage (2) of the penalty score. On the other hand, if you actually do commit a crime you will (ideally) receive the appropriate punishment, and society will be properly protected from you: this is like usage (3) above (but is now equally appropriate if you are a sentient human or a dangerous dog).

For example, in the experiments on forecasting football results that de Finetti ran with his students, they could use method (2) ahead of a game, to help them decide on what probability value to give to a home win; after the game, their degree of success or failure would be measured by method (3). Being sure of the wrong result would lead to the highest penalty, being sure of the right result the lowest; while expressing intermediate uncertainty would be rewarded less when right, but penalised less when wrong.

Usages (1) and (2) for a scoring rule do not relate directly to the criterion of objectivity: (1) is purely formal, while (2) is simply a helpful method for specifying subjective probabilities. However, usage (3) does specifically address this issue: it provides a quantitative measure of how well a probability specification actually performs, in the light of the outcomes that actually occur. Because of the close relationships between all three usages, it is apparent that, so far from having no way to relate subjective probabilities to reality, as is commonly asserted, de Finetti's approach does naturally incorporate this dimension.

The distinction between objectivity and objectivism is fundamental here. The penalty score criterion of success is entirely *objective*, but is not at all *objectivist*. That is, it does not judge the suitability of a probability assessment according to how close it comes to some putative externally existing "objective probability" of the event; and is indeed entirely agnostic (if not downright sceptical) about the existence of such a real-world quantity. Rather it supplies a measure of match or mismatch between probability assessments (which might be entirely subjective, or the outputs of some theoretical calculation) and the *actual outcomes* of the associated events.

II.3   Comparative assessment

If each of a bevy of forecasters has announced her values (no matter how assessed or interpreted) for each of a sequence of events (e.g., the results of various football matches), and the outcomes of these events are then observed, we can evaluate the actual performance of each forecaster by means of her total penalty score over the sequence. This gives us a natural way of comparing the forecasters, and so identifying some as "objectively better" than others – not because their probability forecasts were "closer to the true probabilities", but because they have shown they can perform well in this game against Nature and each other.

However, this simple comparison by means of total penalty score confounds two distinct dimensions of ability: the *substantive* and the *normative*. One probability forecaster might out-perform another because she is more knowledgeable about the events being forecast, and so able to assign valid probabilities closer to 0 or 1, thus obtaining a lower score. This is *substantive* ability. Alternatively, even with identical knowledge, one forecaster may be better able than another to represent that knowledge appropriately in her numerical assessments – for example, unreasonable over-confidence would lead to high penalties. This is the dimension of *normative* ability.

The total Brier score can be separated into two components, one for each of these dimensions. Suppose that, for a set $A_1, \ldots, A_N$ of uncertain events, a probability assessor assigns probability p_i to A_i. We suppose that the values of the p_i are chosen from a finite set $(\pi_1, \ldots, \pi_k)$, each of which may be used a number of times. For each such value π_j, let n_j be the number of events assessed as having probability π_j, and r_j the number of these that in fact occur, so that $\rho_j = r_j / N$ is the proportion, out of those events assessed as having probability π_j, that in fact occur. As illustration, Table 4 displays the values of these quantities when derived from the sequences of probability forecasts and outcomes in Table 3.

The total Brier score over the N events can now be written (Sanders, 1963) as $S = S_1 + S_2$, where $S_1 = \sum_{j=1}^{k} n_j (\rho_j - \pi_j)^2$ and $S_2 = \sum_{j=1}^{k} n_j \rho_j (1 - \rho_j)$. The term S_1 will be small when, for each probability value π_j used, the empirical relative frequency ρ_j, over the set of events associated with that value, is itself close to the issued value π_j. This term thus rewards a forecaster who is normatively good, in the sense that her assigned probability values are mirrored by objective relative frequencies in the real world. In contrast the second term, S_2, does not depend at all on the actual values assessed for the probabilities, but rather on the way the forecaster sorts the events into groups each having a common probability. This penalty term will be smallest when the relative frequencies (ρ_j) of the events in these groups are all close to 0 or 1. Clearly this can only be done with considerable knowledge of what is in fact likely to occur: thus S_2 rewards substantive skill.

II.4   Calibration and refinement

The two dimensions of normative and substantive skill, as measured by S_1 and S_2, are also referred to as *calibration* and *refinement* (or *resolution*), or as *labelling* and *sorting*.

The forecaster has two distinct tasks: first, to sort the various events into subsets she regards as all having essentially the same uncertainty; and, secondly, to label each of these subsets with a numerical measure of that uncertainty. For example, consider two weather forecasters, F_1 and F_2, who have to give, each day, a probability that it will rain on the next day. Suppose that, over a long sequence of days, F_1 always says 0.6, while F_2 (who has a crystal ball) says either 0 or 1, and is always right. Suppose further that the long-term frequency of rain is 60%. Then both forecasters are perfectly calibrated – they have the same degree of normative skill. However the second forecaster, who is correctly able to sort the days according to their raininess, shows more substantive skill. Another forecaster F_3, who always used probabilities of 0 and 1 but is always wrong, has just as much substantive skill as F_2, but is normatively as bad as can be. However, if only we knew how to recalibrate her, her forecasts would be much more valuable than those of F_1.

We might use the refinement component S_2 of the total Brier score to rank forecasters (whether or not well-calibrated) according to their substantive skill. However, if we used some other scoring rule in a similar way we might get a different ordering. In some cases a forecaster F_1 will have a better refinement score than another, F_2, no matter which scoring rule is used: then we can genuinely regard F_1 as "more refined than" F_2.

This happens, essentially, when we can reproduce the forecasts of F_2 from those of F_1 by using additional randomisation, unrelated to the actual events being forecast (de Groot and Fienberg 1983).

II.5   Complete calibration

Another way of evaluating substantive skill is by an extension of the calibration criterion. Consider a sequence of events whose outcomes alternate: 0, 1, 0, 1, 0, 1,.... It should be possible to pick up this pattern and use it to make good forecasts: this, we assume, is done by a forecaster F_2, who forecasts perfectly (and is of course perfectly calibrated). Another forecaster, F_1, always gives forecasts 0.5. He too is perfectly calibrated, over the whole sequence. However, if we restrict attention to (say) the even-numbered events in the sequence, F_2 is still perfectly calibrated, whereas F_1 no longer is. This is because only F_2 has been able to pick up relevant patterns in the data. Similarly, if, say, rainy days follow rainy days, and dry days follow dry days, 80% of the time, a forecaster who always gives probability 80% to tomorrow's weather being the same as today's will be well-calibrated, as will be a forecaster who always says 50%. But only the former will retain calibration for the subsequence consisting of those days following rainy days.

We call a forecaster *completely calibrated* if she remains calibrated for any subsequence of events, picked out in a way that does not depend on the events forecasts (although, as in the last case above, it may depend on the outcomes of previous events). Such a forecaster has essentially discovered, and applied, any patterns that may exist in the data, in order to perfect her forecasts: she has therefore shown essentially perfect substantive skill. It can be shown that, for an indefinitely long sequence of events, two completely calibrated forecasters must end up issuing essentially identical forecasts. These agreed values might then be regarded as *quasi-objective* probabilities, attached to the events in question, and justified by the application of an entirely objective, even though non-objectivist, criterion of empirical validity for putative probabilities.

There are however two important qualifications to the above interpretation. First, there is no general way of ensuring validity according to the complete calibration criterion, and so no way of computing these "objective" probabilities. Secondly, while we have allowed selection of subsequences according to past outcomes, there may be additional relevant information we could use to this effect: e.g. selecting those days following days for which the temperature exceeded 50°C. A forecaster who is completely calibrated according to such an extended set of subsequences will typically perform better than one calibrated on past outcomes alone, and the associated "quasi-objective" probability values will differ: they are thus to be interpreted as relative to an appropriate "information base", and so must still be regarded as representing epistemological, rather than objective, uncertainty.

TABLE 1: BRIER SCORING RULE

A occurs	A does not occur	q
100.0	0.0	.000
95.1	0.1	.025
90.2	0.2	.050
85.6	0.6	.075
81.0	1.0	.100
76.6	1.6	.125
72.2	2.2	.150
68.1	3.1	.175
64.0	4.0	.200
60.1	5.1	.225
56.2	6.2	.250
52.6	7.6	.275
49.0	9.0	.300
45.6	10.6	.325
42.2	12.2	.350
39.1	14.1	.375
36.0	16.0	.400
33.1	18.1	.425
30.2	20.2	.450
27.6	22.6	.475
25.0	25.0	.500
22.6	27.6	.525
20.2	30.2	.550
18.1	33.1	.575
16.0	36.0	.600
14.1	39.1	.625
12.2	42.2	.650
10.6	45.6	.675
9.0	49.0	.700
7.6	52.6	.725
6.2	56.2	.750
5.1	60.1	.775
4.0	64.0	.800
3.1	68.1	.825
2.2	72.2	.850
1.6	76.6	.875
1.0	81.0	.900
0.6	85.6	.925
0.2	90.2	.950

| 0.1 | 95.1 | .975 |
| 0.0 | 100.0 | 1 000 |

You have to choose a row. If the event A happens, you will suffer the penalty given in column 1. If A does not happen, your penalty will be as given by column 2. What is your preferred row? The value of q in column 3 of that row will give your probability p for the event A.

TABLE 2: LOGARITHMIC SCORING RULE

A occurs	A does not occur	q
1151.29	0.00	0.00001
921.03	0.01	0.0001
690.78	0.10	0.001
460.52	1.01	0.010
299.57	5.13	0.050
230.26	10.54	0.100
207.94	13.35	0.125
189.71	16.25	0.150
174.30	19.24	0.175
160.94	22.31	0.200
149.17	25.49	0.225
138.63	28.77	0.250
129.10	32.16	0.275
120.40	35.67	0.300
112.39	39.30	0.325
104.98	43.08	0.350
98.08	47.00	0.375
91.63	51.08	0.400
85.57	55.34	0.425
79.85	59.78	0.450
74.44	64.44	0.475
69.31	69.31	0.500
64.44	74.44	0.525
59.78	79.85	0.550
55.34	85.57	0.575
51.08	91.63	0.600
47.00	98.08	0.625
43.08	104.98	0.650
39.30	112.39	0.675

35.67	120.40	0.700
32.16	129.10	0.725
28.77	138.63	0.750
25.49	149.17	0.775
22.31	160.94	0.800
19.24	174.30	0.825
16.25	189.71	0.850
13.35	207.94	0.875
10.54	230.26	0.900
7.80	259.03	0.925
5.13	299.57	0.950
2.53	368.89	0.975
1.01	460.52	0.990
0.10	690.78	0.999
0.01	921.03	0.9999
0.00	1151.29	0.99999

TABLE 3

Probability p	0.4	0.6	0.3	0.2	0.6	0.3	0.4
Outcome	0	0	1	0	1	0	1

Probability p	0.5	0.6	0.2	0.6	0.4	0.3	0.5
Outcome	1	1	0	1	0	0	1

TABLE 4

Probability π	0.2	0.3	0.4	0.5	0.6
Instances n	2	3	3	2	4
Successes r	0	1	1	2	3
Proportion ρ	0	0.33	0.33	1	0.75

References

Brier, G.W. (1950). "Verification of Forecasts Expressed in Terms of Probability". *Monthly Weather Review* LXXVIII, pp. 1-3.

Dawid, A. P. (1986). "Probability Forecasting". In S. Kotz, N.L. Johnson and C.B. Read (eds.), *Encyclopedia of Statistical Sciences*, vol. 7. Wiley-Interscience, pp. 210-218.

de Finetti, B. (1933). "Sul concetto di probabilità". *Rivista italiana di statistica, economia e finanza* V, pp. 723-47. English edition "On the Probability Concept". In de Finetti (1992), pp. 335-352.

de Finetti, B. (1951). "Recent Suggestions for the Reconciliation of Theories of Probability". In J. Neyman (ed.), *Proceedings of the Second Berkeley Symposium on Mathematical Statistics and Probability*. Berkeley: University of California Press, pp. 217-225.

de Finetti, B. (1962a). "Does it Make Sense to Speak of 'Good Probability Appraisers'?". In I.J. Good *et al.* (eds.), *The Scientist Speculates. An Anthology of Partly-Baked Ideas*. New York: Basic Books, pp. 357-364.

de Finetti, B. (1962b). "Obiettività e oggettività: critica a un miraggio". *La Rivista Trimestrale* I, pp. 343-367.

de Finetti, B. (1968). "Probability: the Subjectivistic Approach". In R. Klibansky (ed.), *La philosophie contemporaine*. Florence: La Nuova Italia, pp. 45-53.

de Finetti, B. (1970a). *Teoria delle probabilità*. Turin: Einaudi. English edition *Theory of Probability*. New York: Wiley, 1975.

de Finetti, B. (1970b). "Logical Foundations and Measurement of Subjective Probability". *Acta Psychologica* XXXIV, pp. 129-145.

de Finetti, B. (1973). "Bayesianism: Its Unifying Role for Both the Foundations and the Applications of Statistics". *Bulletin of the International Statistical Institute, Proceedings of the 39th Session*, pp. 349-68.

de Finetti, B. (1974). "The Value of Studying Subjective Evaluations of Probability", in *The Concept of Probability in Psychological Experiments*, ed. by C.-A. Stäel von Holstein. Dordrecht-Boston: Reidel, pp. 1-14.

de Finetti, B. (1980). "Probabilità". In *Enciclopedia Einaudi*, vol. 10. Turin: Einaudi, pp. 1146-1187.

de Finetti, B. (1992). *Probabilità e induzione (Induction and Probability)*, ed. by P. Monari and D. Cocchi. Bologna: CLUEB (a collection of de Finetti's papers in Italian and English).

de Finetti, B. (1995). *Filosofia della probabilità*, ed. by A. Mura. Milan: Il Saggiatore.

de Groot, M. H. and Fienberg, S. E. (1983). "The Comparison and Evaluation of Forecasters". *The Statistician* XXXII, pp. 12-22.

Galavotti, M.C. (1989). "Anti-realism in the Philosophy of Probability: Bruno de Finetti's Subjectivism". *Erkenntnis* XXXI, pp. 239-261.

Galavotti, M.C. (1995). "F.P. Ramsey and the Notion of 'Chance'". In J. Hintikka and K. Puhl (eds.), *The British Tradition in XXth Century Philosophy*. Wien: Hölder-Pichler-Tempsky, pp. 330-340.

Galavotti, M.C. (1999). "Some Remarks on Objective Chance (F.P. Ramsey, K.R. Popper and N.R. Campbell)". In M.L. Dalla Chiara *et al.* (eds.), *Language, Quantum, Music*. Dordrecht-Boston: Kluwer, pp. 73-82.

Galavotti, M.C. (2001). "Subjectivism, Objectivism and Objectivity in Bruno de Finetti's Bayesianism". In D. Corfield and J. Williamson (eds.), *Foundations of Bayesianism*. Dordrecht-Boston: Kluwer, pp. 161-174.

Galavotti, M.C. (2003). "Harold Jeffreys' Probabilistic Epistemology: Between Logicism and Subjectivism". *British Journal for the Philosophy of Science* LIV, pp. 43-57.

Galavotti, M.C. (2005). *Philosophical Introduction to Probability*. Stanford: CSLI.

Good, I. J. (1952). "Rational Decisions". *Journal of the Royal Statistical Society, Series B,* XIV, pp. 107-114.

Laudan, L. (2006). *Truth, Error and Criminal Law*. Cambridge: Cambridge University Press.

Sanders, F. (1963). "On Subjective Probability Forecasting". *Journal of Applied Meteorology* II, pp. 191-201.

Savage, L.J. (1954). *The Foundations of Statistics*. New York: Wiley. Second edition Dover, 1972.

Savage, L.J. (1971). "Elicitation of Personal Probabilities and Expectations". *Journal of the American Statistical Association* LXVI, pp. 783-801.

Stella, F. (2003). *Giustizia e Modernità*. Milan: Giuffrè (third edition).

7

De Finetti and the Arrow-Pratt Measure of Risk Aversion

ALDO MONTESANO

1 Introduction

Following the formalization of expected utility theory by von Neumann and Morgenstern (1947), various new issues had to be considered in scholars' discussions. One such issue was the question of the measurability of utility (with some confusion between the notion of utility as a representation of preferences, which is ordinal utility, defined up to any strictly increasing transformation, and von Neumann-Morgenstern's utility, which is that particular utility function, defined up to any strictly increasing linear transformation, which allows preferences on lotteries to be represented by a linear functional with regard to probabilities). A second issue was the extension of expected utility theory from the space of lotteries (that is, actions with uncertain outcomes and given probabilities), with which von Neumann-Morgenstern's theory is concerned, to the space of acts (that is, actions with subjective probabilities, which are therefore expressions of individual preferences, as are utilities). Another issue was the empirical validity of expected utility theory; I refer in particular to those criticisms that are emblematically represented by the Allais and Ellsberg paradoxes. There was then a period of intense activity that rendered expected utility theory the essential reference for all analysis and economic applications in the field of choices under uncertainty.

De Finetti had an active and important role in all of these discussions. I will limit myself here to illustrating de Finetti's contribution on the subject of measuring risk aversion, found in the article "Sulla preferibilità"

Bruno de Finetti, Radical Probabilist
Maria Carla Galavotti (ed.)
Copyright © 2009

published in the *Giornale degli Economisti* in 1952. This contribution has remained almost unknown in the literature. Guido Rossi pointed it out to me while I was preparing a lecture (Montesano 1991) that I was to give at the fifth FUR conference at Duke University in 1990, and on that same occasion I pointed it out to John Pratt, who knew nothing of it (and was astonished by it).

In expected utility theory, the decision maker is characterized only by von Neumann-Morgenstern's utility function (or Bernoulli's utility function, as others call it), which has the outcome of actions as its argument. In the standard presentation of the theme of risk aversion, outcomes are sums of money, so that utility is a function of one sole variable. The issue at hand is therefore to examine the link between the type of utility function and the decision maker's preferences, represented by the value that the decision maker gives to the different lotteries. In very general terms, risk aversion (propensity) is linked to a low (high) value being assigned to the generic lottery by the decision maker. Analysis of this consists of deducing the characteristics of the utility function with which this valuation is linked. An initial problem springs from the fact that this characterization is neither immediate nor perfect (for defining a necessary and sufficient condition, that is), if all possible lotteries are considered. De Finetti resolved this problem, as did Arrow and Pratt, with regard to small lotteries, whose possible outcomes differ little amongst themselves. Thus a *local* analysis of risk aversion is proposed. Another problem is found in the definition itself of risk aversion, by which we describe a decision maker who is risk-averse. There are two versions of this, which, however, both lead to the same characterization in the case of expected utility theory. The first of these, from both de Finetti and Pratt, defines a decision maker for whom the value of a lottery is equal to its mean value as risk-neutral, one for whom this value is lower than its mean value as risk-averse, and one for whom it is higher as risk-loving. This definition is linked to a comparative one that defines one decision maker as more risk-averse than another: this is the case where the value of the lottery is lower for the former (in this way, a decision maker is risk-averse if he is more averse to risk than a risk-neutral decision maker). The second version (introduced subsequently) first gives a definition of when one lottery is riskier than another: Rothschild and Stiglitz (1970), for example, identify lotteries obtained from the lottery under examination through a mean preserving spread as being riskier (so, for example, lowering the probability of one outcome and raising the probability of two outcomes in order to leave the mean value unchanged). Next, a decision maker who prefers less risky lotteries to more risky ones is described as risk-averse. These two versions give rise to two different definitions of risk aversion. I call the first "risk aversion", and the second "aversion to increasing risk"; in the literature they are often referred to as "weak risk aversion" and "strong risk aversion", because the second implies the first. In the case of expected utility theory, both depend on the concavity of utility function. That is, if, and only if, utility function is concave, the risk premium (the difference between the mean value and the certainty equivalent of a lottery) is non-negative for every lottery, and less risky lotteries are (weakly) preferred to more risky ones.

The observation that the concavity of utility function has a role in risk aversion can be dated back to Bernoulli, who had introduced a logarithmic function to obtain a finite certainty equivalent for a lottery, such as that of St. Petersburg, with an infinite mean value. The problem, therefore, is not so much associating risk aversion to concavity, as finding which concavity index is a measure of risk aversion. De Finetti (1952) was the first to do this. The index was then found by Arrow (1965) and Pratt (1964), and is universally known as the Arrow-Pratt index. Pratt made an extremely valuable analysis of risk aversion. De Finetti's analysis is very concise, no longer than a page of a journal (pp. 700-701), is elliptical, and contains some misprints. However, it is very rich in content. The introduction of the risk aversion index is discussed in four different ways (following two put forward by Pratt and one by Arrow). Some explanation is called for.

2 De Finetti's analysis

De Finetti's concise analysis of risk aversion is based on expected utility theory and is local since it considers behavior towards risk with regard to lotteries whose outcomes differ little amongst themselves. He writes (1952: 700) that "probabilistic behavior around a certain value x depends on the degree of relative convexity of u therewith, relative convexity meaning the ratio $-u''/2u'$ between the second and first derivatives". In this respect he uses the function $\lambda(x) = -2u'(x)/u''(x)$, which has the physical dimension of the argument of utility function (that is, that of money for monetary lotteries).

De Finetti analyzes the degree of risk aversion according to four relationships, which it is useful to examine in detail.[1]

a) De Finetti sets out the first relationship in this way: "the risk of gaining or losing h with equal chances ($\pm h$ with prob. ½ and ½) is equivalent to a sure loss h^2/λ". That is, de Finetti introduces the lottery $\ell(h) = (h, \frac{1}{2} ; -h, \frac{1}{2})$ and, denoting the decision maker's wealth with x, considers the certainty equivalent $CE(x, \ell(h))$ of this lottery, and shows that $CE(x, \ell(h)) = -h^2/\lambda(x)$ for $h \to 0$. This relationship can be demonstrated bearing in mind that

$$u(x + CE(x, \ell(h))) = \frac{1}{2}u(x+h) + \frac{1}{2}u(x-h)$$

Thus, deriving twice with respect to h,

$$u'(x + CE(x, \ell(h)))\frac{\partial CE(x, \ell(h))}{\partial h} = \frac{1}{2}u'(x+h) - \frac{1}{2}u'(x-h)$$

$$u''(x+CE(x,\ell(h)))\left(\frac{\partial CE(x,\ell(h))}{\partial h}\right)^2 + u'(x+CE(x,\ell(h)))\frac{\partial^2 CE(x,\ell(h))}{\partial h^2} =$$

$$= \frac{1}{2}u''(x+h) + \frac{1}{2}u''(x-h)$$

So, at $h = 0$, we find that $CE(x, \ell(0)) = 0$, $\left.\dfrac{\partial CE(x, \ell(h))}{\partial h}\right|_{h=0} = 0$ and

$\left.\dfrac{\partial^2 CE(x, \ell(h))}{\partial h^2}\right|_{h=0} = \dfrac{u''(x)}{u'(x)}$. Therefore, in the neighborhood of $h = 0$,[2]

$$CE(x, \ell(h)) = -h^2/\lambda(x) + o(h^2)$$

Taking into account that the mean value of the lottery $\ell(h)$ is $EV(\ell(h)) = 0$, de Finetti's first relationship can be expressed saying that

---

[1] De Finetti does not use the term "risk aversion" in his presentation of these relationships, but rather uses generic terms, such as "probabilistic behavior" and "advantageous, disadvantageous, indifferent bets". However, he does use the term "degree of risk aversion" to denote what he calls "relative convexity", that is u''/u', immediately afterwards (p. 701).

[2] Here onwards it is assumed that, when they occur, functions are expandable in Taylor series. The presence of terms of smaller order than h^n is indicated by $o(h^n)$, and $O(h^n)$ is used to indicate the presence of terms of an order at most equal to h^n.

the risk premium (a sure loss, in de Finetti's terminology) $EV(\ell(h)) - CE(x, \ell(h))$ is, for $h \to 0$, proportional to h^2, with proportionality coefficient equal to $1/\lambda(x)$, which is, therefore, a measure of the decision maker's risk aversion.[3]

b) The second relationship states that to render the bet indifferent "the probability of winning must be greater than that of losing by $d = h/\lambda$ (so, gain $\pm h$ with respective probability $(1\pm d)/2)$". That is, de Finetti introduces the lottery $\ell(d) = (h, (1+d)/2 ; -h, (1-d)/2)$, considers its certainty equivalent $CE(x, \ell(d))$, seeks the value of d at which $CE(x, \ell(d)) = 0$, and finds that $d(x, h) = h/\lambda(x)$ for $h \to 0$. This relationship can be demonstrated remembering that

$$u(x) = u(x + CE(x, \ell(d))) = u(x + h)\frac{1 + d(x, h)}{2} + u(x - h)\frac{1 - d(x, h)}{2}$$

Thus, deriving twice with respect to h,

$$0 = u'(x + h)\frac{1 + d(x, h)}{2} - u'(x - h)\frac{1 - d(x, h)}{2} +$$

$$+ \frac{1}{2}(u(x + h) - u(x - h))\frac{\partial d(x, h)}{\partial h}$$

$$0 = u''(x + h)\frac{1 + d(x, h)}{2} + u''(x - h)\frac{1 - d(x, h)}{2} + (u'(x + h) + u'(x - h))\frac{\partial d(x, h)}{\partial h} +$$

$$+ \frac{1}{2}(u(x + h) - u(x - h))\frac{\partial^2 d(x, h)}{\partial h^2}$$

So, at $h = 0$, we have $d(x, 0) = 0$ and $\left.\frac{\partial d(x, h)}{\partial h}\right|_{h=0} = -\frac{1}{2}\frac{u''(x)}{u'(x)}$. Therefore, in the neighborhood of $h = 0$,

$$d(x, h) = h/\lambda(x) + o(h)$$

---

[3] De Finetti gives this proof in a footnote, which, however, appears to be confused (an explicit definition of the symbol d is lacking, there is a misprint, and two distinct relationships appear to be united). De Finetti uses d to indicate the risk premium, that is $u(x - h) + u(x + h) = 2u(x - d)$. Since $u(x + h) = u + hu' + h^2(u'' + \varepsilon)/2$, $u(x - h) = u - hu' + h^2(u'' + \varepsilon)/2$ and $u(x - d) = = u - du' + d^2(u'' + \varepsilon)/2$, where u, u', u'' indicate $u(x)$, $u'(x)$, $u''(x)$ and ε is a function of h tending to zero with h, then $u(x - h) + u(x + h) = 2u + h^2(u'' + \varepsilon)$ and $2u(x - d) = = 2u - 2du' + d^2(u'' + \varepsilon)$, so that $h^2(u'' + \varepsilon) = -2du' + d^2(u'' + \varepsilon)$ (in de Finetti's article $2du'$ is found in place of $-2du'$) and so, overlooking the last addendum and ε, we obtain $h^2u'' = -2du'$, that is $d = h^2/\lambda$.

This second relationship put forward by de Finetti can be expressed by saying that the probability premium d (to render the lottery $(x+h,(1+d)/2\ ;x-h,(1-d)/2)$ acceptable to the decision maker) is proportional to h for $h \to 0$, with proportionality coefficient equal to $1/\lambda(x)$, which is therefore a measure of the decision maker's risk aversion, even under this aspect.

c) The third relationship that de Finetti proposes states that "the price of a gain h (or a loss if $h < 0$) with probability p equals $ph(1+h/\lambda)$". That is, de Finetti introduces the lottery $\ell(h)=(h,p\ ;0,1-p)$ and considers its certainty equivalent $CE(x,\ell(h))$, which is the price of gain h with probability p. De Finetti writes that this price is equal to $ph(1+h/\lambda)$. There is presumably a misprint in this expression, whose correct formula is $ph(1-(1-p)h/\lambda)$. (Amongst other things, the price of the gain must be equal to the gain for $p = 1$). This relationship can be demonstrated using $g(x, h)$ to denote the price of the gain and bearing in mind that

$$u(x+g(x,h)) = u(x+CE(x,\ell(h))) = u(x+h)p+u(x)(1-p)$$

Deriving twice with respect to h, we obtain

$$u'(x+g(x,h))\frac{\partial g(x,h)}{\partial h} = u'(x+h)p$$

$$u''(x+g(x,h))\left(\frac{\partial g(x,h)}{\partial h}\right)^2 + u'(x+g(x,h))\frac{\partial^2 g(x,h)}{\partial h^2} = u''(x+h)p$$

So, at $h = 0$, we have $g(0) = 0$, $\left.\dfrac{\partial g(x,h)}{\partial h}\right|_{h=0} = p$ and $\left.\dfrac{\partial^2 g(x,h)}{\partial h^2}\right|_{h=0} =$

$= p(1-p)u''(x)/u'(x)$. Therefore, in the neighborhood of $h=0$,

$$g(h) = ph - p(1-p)h^2/\lambda(x) + o(h^2)$$

De Finetti's third relationship can be expressed by saying that the price that the decision maker is willing to pay to achieve gain h with probability p, is proportional to the mean gain hp, with proportionality coefficient equal to $1-(1-p)h/\lambda(x)$, which is as much smaller as the decision maker's risk aversion $1/\lambda(x)$ is larger.

d) In the fourth relationship, "in general a bet carrying a random gain X (with X certainly included between $\pm h$) proves to be advantageous, disadvantageous, or indifferent according to whether σ^2 is smaller, larger or equal to $-\lambda m$ (m and σ, expected value and standard deviation of X)". This proposition (which contains a misprint since the correct indication is λm rather than $-\lambda m$) can be demonstrated in the following manner. Considering the lottery $X = (h_i, p_i)_{i=1}^{n}$, with $m = \sum_{i=1}^{n} h_i p_i$ and $\sigma^2 = \sum_{i=1}^{n} (h_i - m)^2 p_i$, and indicating its certainty equivalent with $CE(x, X)$, we have $u(x + CE(x, X)) = \sum_{i=1}^{n} u(x + h_i)p_i$. Thus, deriving twice with respect to $(h_i)_{i=1}^{n}$, we find

$$u'(x + CE(x, X))\frac{\partial CE(x, X)}{\partial h_i} = u'(x + h_i)p_i, \qquad i = 1, ..., n$$

$$u'(x + CE(x, X))\frac{\partial^2 CE(x, X)}{\partial h_i \, \partial h_j} + u''(x + CE(x, X))\frac{\partial CE(x, X)}{\partial h_i}\frac{\partial CE(x, X)}{\partial h_j} =$$

$$= \begin{cases} 0 & \text{if } j \neq i \\ u''(x + h_i)p_i & \text{if } j = i \end{cases} \qquad i, j = 1, ..., n$$

So, if $h_i = m$ for $i = 1, ..., n$, that is if $X = m$, we obtain

$$CE(x, X)\big|_{X=m} = m$$

$$\frac{\partial CE(x, X)}{\partial h_i}\bigg|_{X=m} = p_i, \qquad i = 1, ..., n$$

$$\frac{\partial^2 CE(x, X)}{\partial h_i \, \partial h_j}\bigg|_{X=m} = \begin{cases} -\dfrac{u''(x + m)}{u'(x + m)}p_i p_j & \text{if } j \neq i \\ -\dfrac{u''(x + m)}{u'(x + m)}(p_i^2 - p_i) & \text{if } j = i \end{cases} \qquad i, j = 1, ..., n$$

Hence, around the vector with all components equal to m,

$$CE(x, X) = m + \frac{1}{2}\frac{u''(x + m)}{u'(x + m)}\sigma^2 + o(\sigma^2)$$

and therefore

$$CE(x, X) = m - \frac{\sigma^2}{\lambda(x)} + o(\sigma^2)$$

De Finetti's fourth relationship establishes that willingness to accept a bet depends on $\lambda m - \sigma^2$, where m and σ are characteristics of the bet and λ is characteristic of the decision maker. Therefore λ denotes propensity to accept bets, and so, in this sense, $1/\lambda$ measures risk aversion.

3 Arrow and Pratt's analysis of risk aversion

With reference to monetary utility function $u(x)$, Pratt (1964) defines the local risk aversion

$$r(x) = -u''(x)/u'(x)$$

and explains this in two ways, which correspond to de Finetti's fourth and second relationships respectively.

a) Pratt defines risk $\tilde{z}$ as a monetary lottery whose outcomes increase (or decrease) wealth x. The risk premium $\pi(x, \tilde{z})$ (which is the difference between the mean value of lottery $\tilde{z}$ and its certainty equivalent) is that sum of money that renders the risk $\tilde{z}$ equivalent to the sure sum $EV(\tilde{z}) - \pi(x, \tilde{z})$, i.e. the sum that satisfies the relationship

$$u(x + EV(\tilde{z}) - \pi(x, \tilde{z})) = EU(x + \tilde{z})$$

Local risk aversion is introduced assuming that risk $\tilde{z}$ has small variance σ_z^2, in the sense that the analysis is conducted for $\sigma_z^2 \to 0$ (assuming, moreover, that risk $\tilde{z}$ has a null mean value, that is $EV(\tilde{z}) = 0$, without loss of generality, and that the third absolute central moment is of a smaller order than σ_z^2). Then, taking the preceding relationship into account and expanding $u(x - \pi)$ in series around $\pi = 0$, so that $u(x - \pi) = u(x) - u'(x)\pi + O(\pi^2)$ and expanding $EU(x + \tilde{z})$ for $\sigma_z^2 \to 0$, so that $EU(x + \tilde{z}) = u(x) + \dfrac{1}{2}u''(x)\sigma_z^2 + o(\sigma_z^2)$, we obtain

$$\pi(x, \tilde{z}) = -\frac{1}{2}\frac{u''(x)}{u'(x)}\sigma_z^2 + o(\sigma_z^2) = \frac{1}{2}r(x)\sigma_z^2 + o(\sigma_z^2)$$

The first interpretation of $r(x)$ therefore consists of the fact that the risk premium depends on $r(x)$. (Pratt also points out that the risk premium is proportional to the variance of the risk, not to its standard deviation, thus producing a relationship of the second order. This is relevant with respect to the analysis produced many years later, when, abandoning expected utility theory, a risk aversion of the first order also appeared).

b) For the second interpretation, Pratt considers a risk with only two possible opposing outcomes, $\tilde{z} = (h, p_{+h}; -h, p_{-h})$, where h and $-h$ indicate the two outcomes and p_{+h} and p_{-h} the corresponding probabilities. The probability premium is the difference between these two probabilities that makes it indifferent for the decision maker to run the risk or not. That is, given that $p(x, h) = p_{+h} - p_{-h}$, since $p_{+h} + p_{-h} = 1$, we have $p_{+h} = (1 + p(x, h))/2$ and $p_{-h} = (1 - p(x, h))/2$, and the probability premium results from the relationship

$$u(x) = \frac{1}{2}(1 + p(x, h))u(x + h) + \frac{1}{2}(1 - p(x, h))u(x - h)$$

Expanding $u(x + h)$ and $u(x - h)$ in series around $h = 0$, we find

$$u(x) = u(x) + p(x, h)u'(x)h + \frac{1}{2}u''(x)h^2 + O(h^3)$$

from which

$$p(x, h) = \frac{1}{2}r(x)h + O(h^2)$$

So, the probability premium depends on $r(x)$.

Apart from defining local risk aversion, Pratt also introduces comparative risk aversion, both local and global, establishing some notable equivalence relationships on this matter.

Comparative risk aversion is defined by stating that the utility function u_1 denotes, at point x, a risk aversion that is greater than the function u_2 if $r_1(x) > r_2(x)$. This property is global if the inequality is satisfied at every x.[4] Pratt demonstrates the following relationships as equivalent:

(a) $r_1(x) \geq r_2(x)$

(b) $\pi_1(x, \tilde{z}) \geq \pi_2(x, \tilde{z})$

(c) $p_1(x, h) \geq p_2(x, h)$

(d) $u_1(u_2^{-1}(t))$ is a concave function of t

---

[4] The property is global in the sense that $CE_1(\ell) \leq CE_2(\ell)$ for every ℓ if and only if $r_1(x) \geq r_2(x)$ for every x, and similarly that $CE(\ell) \leq EV(\ell)$ for every ℓ if and only if $r(x) \geq 0$ for every x, not in the sense that it is necessary that $r(x) \geq 0$ for every x in order to have $CE(\ell) \leq EV(\ell)$ for a specific ℓ. That is, there can be lotteries with $CE(\ell) < EV(\ell)$ or $CE(\ell) > EV(\ell)$ even if the sign of $r(x)$ is not always positive or negative, but depends on x.

(e) $\dfrac{u_1(y)-u_1(x)}{u_1(w)-u_1(v)} \leq \dfrac{u_2(y)-u_2(x)}{u_2(w)-u_2(v)}$ for $v < w \leq x < y$

Expression (d) shows that u_1 is a concave transformation of u_2, or $u_1(x) = f(u_2(x))$, where $f(.)$ is a concave function (taking $t = u_2(x)$, we in fact find $u_1(u_2^{-1}(t)) = u_1(x) = f(u_2(x)) = f(t)$).

The other two aspects that Pratt analyzes in his article regard the theme of increasing or decreasing aversion to risk with respect to wealth, that is, the analysis of the function $r(x)$, and the relative (or proportional) risk aversion, measured by the function $r*(x) = xr(x)$ and pertinent to lotteries of the type $x\,\tilde{z}$ (instead of $x+\tilde{z}$). These aspects are not, however, relevant to the ends of the present paper.

Arrow's contribution (1965: 32-35) to measuring risk aversion is less elaborate. He introduces the measures of absolute risk aversion $R_A(Y) = -U''(Y)/U'(Y)$ and relative risk aversion $R_R(Y) = -YU''(Y)/U'(Y)$, taking into account that the function $U''(Y)$ is affected by a linear transformation of the utility function, while $R_A(Y)$ and $R_R(Y)$ are unaffected. He justifies $R_A(Y)$ considering the lottery $\ell = (Y+h, p;\ Y-h, 1-p)$ with certainty equivalent $CE(\ell) = Y$, similarly to de Finetti's second relationship. He finds that

$$p(Y,h) = \frac{1}{2} + \frac{R_A(Y)}{4} h + o(h)$$

that is, that absolute risk aversion measures how favorable a lottery must be to be accepted by the individual. He similarly explains relative risk aversion with the lottery $\ell = (Y+nY, p;\ Y-nY, 1-p)$, finding that

$$p(Y,nY) = \frac{1}{2} + \frac{R_R(Y)}{4} n + o(n)$$

4 On the evolution of the concept of risk aversion and uncertainty aversion

The introduction of the Arrow-Pratt index of risk aversion has fostered new research in many directions. With reference to expected utility theory I shall mention only two examples of this: the analysis of risk aversion's dependence on wealth, that is of r on x (where $r(x)$ is the Arrow-

Pratt index), with particular regard to financial choices, already introduced by Pratt; and the extension of the measure of risk aversion to the case in which utility is a function of several variables (rather than only of money), for example in Khilstrom and Mirman (1974).

However, more interesting analyses were produced with the introduction of theories different from expected utility theory. In these, risk aversion no longer depends, or does not only depend, on the utility function defined on a set of outcomes, but on other functions characterizing the behavior of the decision maker. Moreover, when acts are considered (for which there are no assigned probabilities), it is not even always hypothesized that the subjective probabilities are additive, or that there is a single order of probability. So, a notion of aversion is born that is not concerned with risk (that is, with the fact that diverse outcomes are possible), but rather with uncertainty, or ambiguity, that is with the fact that chances can be more or less known.

The introduction of more general models than the expected utility ones also brought about the realization that there can be risk aversion of a greater order than that evidenced by the Arrow-Pratt index. In short, with expected utility theory the Taylor expansion of the risk premium $EV(\ell) - CE(\ell)$ presents only terms of the second order (proportional to $-u''/u'$) or of orders greater than the second. With more general theories that assume nothing aside from the existence of a certainty equivalent function $CE(\ell)$, or that modify expected utility theory to a limited extent, as in *rank dependent expected utility*, there is also a term of the first order (see Montesano 1985 and 1988; Hilton 1988; and Segal and Spivak 1990, as indicated by Machina 2001: 232), which is thus predominant with respect to the Arrow-Pratt term, at least for small lotteries. Moreover, risk aversion and increasing risk aversion prove to be dependent on different characteristics of the decision maker's preferences, and increasing risk aversion also depends on how the greater riskiness of one lottery versus another is defined.

The presence of uncertainty aversion renders the picture much more complicated, as there can be both risk aversion/propensity and uncertainty aversion/propensity together. Moreover, there are transformations, like the probabilistic mixtures, that render acts less uncertain and more risky in the meantime.[5] We can, in any case, proceed in the analysis of

---

[5] It is natural to define a decision maker as averse to increasing uncertainty if he prefers less uncertain acts to more uncertain ones. A relation of "lower uncertainty" can be introduced, remembering that uncertainty consists of not knowing the probabilities, by means

risk and uncertainty aversion according to de Finetti and Pratt's type of reasoning; introducing a risk and uncertainty premium and seeking the conditions that determine its sign.

5 Conclusions

There is no substantial difference between the measure of risk aversion proposed by de Finetti and the measure proposed by Arrow and Pratt. Logically one should talk about a de Finetti index, or a de Finetti-Arrow-Pratt index. In any case, one cannot fail to note how de Finetti anticipated subsequent analyses. Was he ahead of his time? It would seem not. Measuring risk aversion was a ripe fruit when de Finetti harvested it. He merely lived in a place and wrote in a language that were peripheral with regard to the centre of the scientific world.

References

Arrow, K. J. (1965). *Aspects of the Theory of Risk-Bearing*. Helsinki: Yrjö Jahnssonin Säätiö.

de Finetti, B. (1952). "Sulla preferibilità". *Giornale degli Economisti* NS 11, pp. 685-709.

Hilton, R. W. (1988). "Risk Attitude under Two Alternative Theories of Choice under Risk". *Journal of Economic Behavior and Organization* 9, pp. 119-36.

Khilstrom, R. E. and Mirman, L. J. (1974). "Risk Aversion with Many Commodities". *Journal of Economic Theory* 8, pp. 361-88.

Machina, M. J. (2001). "Payoffs Kinks in Preferences over Lotteries". *Journal of Risk and Uncertainty* 23, pp. 207-60.

Montesano, A. (1985). "The Ordinal Utility under Uncertainty and the Measure of Risk Aversion in Terms of Preferences". *Theory and Decision* 18, pp. 73-85.

Montesano, A. (1988). "The Risk Aversion Measure without the Independence Axiom". *Theory and Decision* 24, pp. 269-88.

---

of a probabilistic mixture. In this way, the uncertainty of act $\lambda a_a \oplus (1-\lambda) a_b$, obtained as the probabilistic mixture of the two acts a_a and a_b, is not larger to the most uncertain of the two acts. Thus aversion to increasing uncertainty (with neutrality towards increasing risk) requires that $CE(\lambda a_a \oplus (1-\lambda) a_b) \geq \min\{CE(a_a), CE(a_b)\}$. The following example illustrates the case in which a probabilistic mixture succeeds in eliminating uncertainty: with $a_a = (1, E_1; 0, E_2)$, $a_b = (0, E_1; 1, E_2)$ and $\lambda = \frac{1}{2}$, we have $\lambda a_a \oplus (1-\lambda) a_b = ((1, \frac{1}{2}; 0, \frac{1}{2}), E_1; (0, \frac{1}{2}; 1, \frac{1}{2}), E_2) = (1, \frac{1}{2}; 0, \frac{1}{2})$. On the other hand, a possible definition of "greater risk" is that which associates risk with the presence of probability nodes, so the act $\lambda a_a \oplus (1-\lambda) a_b$ has a riskiness not inferior to the least risky of the two acts, thus aversion to increasing risk (with neutrality towards increasing uncertainty) requires $CE(\lambda a_a \oplus (1-\lambda) a_b) \leq \max\{CE(a_a), CE(a_b)\}$.

Montesano, A. (1991). "Measures of Risk Aversion with Expected and None-xpected Utility". *Journal of Risk and Uncertainty* 4, pp. 271-83.

Montesano, A. (1999). "Risk and Uncertainty Aversion on Certainty Equivalent Functions". In *Beliefs, Interactions and Preferences in Decision Making*, ed. by M. J. Machina and B. Munier. Dordrecht: Kluwer, pp. 23-52.

von Neumann, J. and Morgenstern, O. (1947, 2nd ed.). *Theory of Games and Economic Behavior*. Princeton: Princeton University Press.

Pratt, J. W. (1964). "Risk Aversion in the Small and in the Large". *Econometrica* 32, pp. 122-36.

Rothschild, M. and Stiglitz. J. E. (1970). "Increasing Risk: I. A Definition". *Journal of Economic Theory* 2, pp. 225-43.

Segal, U. and Spivak, A. (1990). "First Order versus Second Order Risk Aversion". *Journal of Economic Theory* 51, pp. 111-25.

8

The Feasibility of Normative Structures

ROBERTO SCAZZIERI[*]

1 Introduction

Normative structures are central to Bruno de Finetti's contribution to economics. In this paper, I shall argue that de Finetti's theory of normative structures points to important features of decision theory, provides an interesting bridge between decision theory and ethical theory, and suggests a sophisticated analytical foundation for de Finetti's theory of political economy.

The paper is organized as follows. Section II examines the building blocks of de Finetti's theory of normative structures and discusses de Finetti's idea that normative structures associated with multiple objectives are central to human decision making. This section briefly reconstructs de Finetti's theory of optimum problems and its complex relationship to the general theory of maximization. Section III discusses the way in which, according to de Finetti, the theory of optimum decisions suggests a treatment of "circumscribed" rationality rooted in the internal structure of rational decisions once it is acknowledged that rational decisions are almost regularly associated with multiple, and partially conflicting, objectives. Section IV goes back to de Finetti's analysis of the relationship between

---

[*] This paper is the outcome of research carried out within the framework of the Focus Group "Dynamics of Human Knowledge" at the Institute of Advanced Study of the University of Bologna. I acknowledge my debt to the Bigiavi Library of Economics, University of Bologna, for allowing unlimited access to its resources. The original draft of this paper was presented at the workshop "Bruno de Finetti, Radical Probabilist" (University of Bologna, 26-28 October 2006). I am grateful to workshop participants for stimulating discussion although, of course, the usual caveat applies.

Bruno de Finetti, Radical Probabilist
Maria Carla Galavotti (ed.)

optimum theory and the theory of maximization, and discusses in that light de Finetti's views on competitive equilibria and their optimal properties. Section V presents the fundamental structure of de Finetti's theory of pure preference, and highlights the relationship between that general framework and a variety of interpretations and applications. In this context, it is emphasized that the theory of optimum decisions with multiple objectives is formally analogue to the theory of rational decisions under uncertainty (a link suggested by Bruno de Finetti himself). This implies that the formal analysis of preference structures may be associated with the consideration of internal constraints, due either to the relationship among objectives or to the lack of knowledge about possible outcomes. Section VI brings the paper to close by considering the relationship between the theory of pure preference, its various "circumscriptions" (due to context), and de Finetti's view of economic theory as a theory of normative structures and feasible design. It is argued that de Finetti's theory of feasible normative structures presupposes awareness of context, and ability to adjust partial objectives in the light of an open and pragmatic approach to the heterogeneity of goals (both for any given decision maker and across different decision makers). "Ultimate" goals are seen as the outcome of a complex hierarchical structure of partial objectives. In this way, normative structures suggest the possibility of an economic theory of pure preference centred upon the critical discussion of goals and the attempt to conceive feasible institutional arrangements consistent with the postulated objectives.

2 The variety of optimum decisions

Bruno de Finetti was interested in the formal analysis of normative structures since his critical reconstruction of Pareto's theory of optimum decisions (de Finetti 1935, 1936, 1937a, 1937b). A characteristic premise of such a reconstruction is the view that some of Pareto's conclusions may be unwarranted but that it would be wrong "to unleash an attack against the whole conception of Pareto, by rejecting his approach and method of inquiry. For it seems to me that only such an approach and method are conducive to criticism and productive analysis of the conception and results of Pareto himself " (de Finetti 1969: 64). The starting point of de Finetti's reconstruction of optimum theory is the idea that optimum solutions are not unique. De Finetti's original argument runs as follows:

Let us assume that the ophelimities [utility levels in Pareto's terminology] $\Phi_1 = a_1$, $\Phi_2 = a_2$, ..., $\Phi_{n-1} = a_{n-1}$ for n-1 individuals are fixed. On the variety [in the mathematical sense] that is thus defined, Φ_n will have a maximum value. We would therefore have at least one optimum point on it. (de Finetti 1936: 319)

After acknowledging the above property, de Finetti proceeds to discuss the cardinality of the whole set of optimum points (as defined above):

There are at least ∞^{n-1} such optimum points; they actually bring about a variety of n-1 dimensions if there is a single optimum for any given variety $\Phi_1 =$ constant , $\Phi_2 =$ constant , ..., $\Phi_{n-1} =$ constant, and such uniqueness theorem holds under assumptions that are natural enough. However, it is not difficult to think of examples in which the optimum points bring about a variety of more than n-1 dimensions. (de Finetti 1936: *ibidem*)

In other words, there are infinite optimum points, which may be identified by considering a space of dimension n-1. As we vary the utility levels for any one of individuals $1,2,...,$ n-1, the maximum value of Φ_n will also change, and the optimum point will be different. If we assume that, for any given set of utility levels $\Phi_1 =$ constant, $\Phi_2 =$ constant, ...,$\Phi_{n-1} =$ constant, the maximum value of utility Φ_n is unambiguously determined, there will be a single optimum for any given collection $\Phi_1 =$ constant, $\Phi_2 =$ constant, ...,$\Phi_{n-1} =$ constant. However, the utility of any one individual is not always unambiguously determined. For example, there may be individuals whose utility is in fact decomposable into a certain number of distinct utility functions. In this case, there will be more than a single optimum for any given collection $\Phi_1 =$ constant, $\Phi_2 =$ constant, ...,$\Phi_{n-1} =$ constant, and it would be reasonable to conjecture that 'the optimum points bring about a variety of more than n-1 dimensions' (de Finetti: *ibidem*). An interesting implication is that, for any given collection of utility functions *1, 2, 3*, any optimum point for utility functions *1, 2* is also an optimum point relative to *1,2,3*. This is because an optimum point on the *1,2* variety presupposes, say, *1* constant and *2* maximum, or vice versa. As we switch to the *1,2,3* variety, we discover that an optimum point presupposes, for example, *1* and *3* constant and *2* maximum. If we set *3=0*, an optimum point on the *1,2* variety will also be an optimum on the *1,2,3* variety. This means that it would generally be possible "to find points, lines, etc. of 'optimum' ... in a more straightforward and easy way before attaining a complete solution of the problem" (de Finetti

1937a: 63). In general terms, the set omega of the ∞^{n-1} optimum points strictly includes the set of the ∞^{m-1} optimum points if $m < n$. De Finetti illustrates this proposition by examining the case in which the variety of optimum dimensions is increased, say from m to n, by considering the linear combinations associated with a finite set of "primitive" objective functions:

> [If] a linear combination with positive coefficients (which we shall briefly call linear positive combination of the Φ_h, such as $\Phi = \sum \rho_h \Phi_h$, $\rho_h \geq 0$) has its absolute maximum at a point P, such a point is necessarily also an optimum point. For, if there would be a point Q such that all functions Φ_h would take greater values than at point P, then also $\Phi = \sum \rho_h \Phi_h$ would take a greater value at Q than at P, which would be against our assumption. (de Finetti 1937a: 63)

A simple way to look at de Finetti's argument, is to consider the case of two objective functions (or utility functions) Φ_1 and Φ_2. Let the values of the two functions be represented in the figure below:

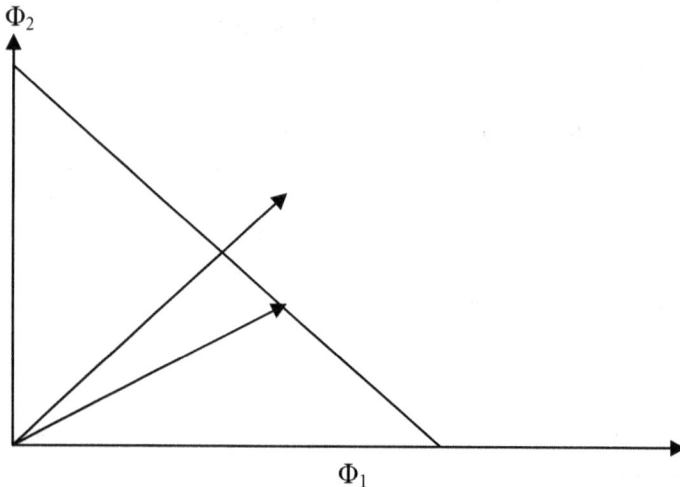

Figure 1. *Positive linear combinations of utility functions: from maximum values to optimum points*

Let us consider the positive linear combination $P = \sum \rho_1 \Phi_1 + \rho_2 \Phi_2$ (as represented by the end point of the bottom arrow inside figure *1*). At point $\Phi_1 max$, function Φ_1 has a maximum value for the value of $\Phi_2 = 0$. At point $\Phi_2 max$, function Φ_2 has a maximum value for the value of $\Phi_1 =$

0. Any point on the segment Φ_1 Φ_2 is a maximum of the positive linear combination $\Phi = \sum \rho_1 \Phi_{1+} \rho_2 \Phi_2$ for it would be impossible to raise the value of Φ_1 without reducing the value of Φ_2 (or vice versa). This point, as de Finetti points out, is of necessity (that is, by definition) also an optimum point. Let us now consider another point $Q = \sum \rho_1 \cdot \Phi_{1+} \rho_2 \cdot \Phi_2$ such that $\rho_1 \neq \rho_1$ and $\rho_2 \neq \rho_2$ (this point is represented by the end point of the top arrow inside figure 1). If both functions Φ_1 and Φ_2 take greater values than at point P, then also the positive linear combination $\Phi' = Q = \sum \rho_1 \cdot \Phi_1 + \rho_2 \cdot \Phi_2$ would take a greater value at Q than at P, which would be against the assumption that the linear combination has an absolute maximum at P. The logical substance of de Finetti's argument is that we do not need to consider the complete variety of optimum dimensions in order to identify a finite set of optimum points. In particular, we do not need to know the full range of optimum points associated with all positive linear combinations Φ_h of functions Φ_1 and Φ_2 as long as we know the optimum points associated with utility functions Φ_1 and Φ_2. This property of the topological space of optimum points entails a remarkable cognitive economy, as some optimum points may be identified by considering only a finite region of the whole space (see above).

De Finetti maintains that, under suitable assumptions, "also the inverse conclusion holds: if P is an optimum point, it is also the point associated with the absolute maximum of one of the positive linear combinations $\Phi = \sum \rho_h \Phi_h$" (de Finetti 1937a: 66) and that it is possible to show "the continuity of the above correspondence" in the sense that, "if we slightly change $\rho_1, \rho_2, ..., \rho_h$ also the absolute maximum for $\Phi = \sum \rho_h \Phi_h$ can only change but a little" (de Finetti: *ibidem*). Figure 2 may be helpful to interpret this "inverse conclusion".

Let us consider the positive linear combination $P = \sum \rho_1 \Phi_{1+} \rho_2 \Phi_2$. The intersection of the locus of all positive linear combinations with the horizontal axis is itself a value of the linear combination if we assume $\rho_2 = 0$. This point is obviously a maximum of this particular linear combination (that is, if we assign zero weight to function Φ_2) as any other value of it would be associated with a lower value of the utility function Φ_1. It is easy to see that a slight change of weights ρ_1, ρ_2 will be associated with only a slight change in the maximum value of the linear combination.

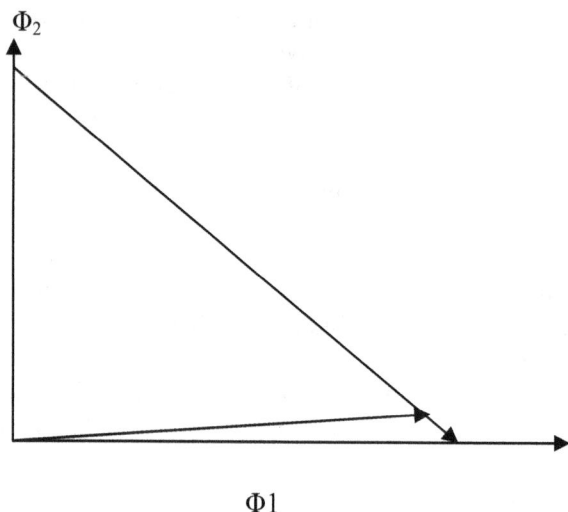

Figure 2. *Positive linear combinations of utility functions: from optimum points to maximum values*

For example, if ρ_2 rises to a small positive value $\varepsilon > 0$, the maximum of the new linear combination will be associated with a point on the locus close to the intersection (as shown by the arrow from the origin). In this case, the rise of ρ_2 to a small positive value will be associated with a decrease of ρ_1, as otherwise the original position (the intersection) would not be an optimum point.

De Finetti establishes the relationship between maximum and optimum points by considering an optimum space of dimension *n-1* for any given set of *n* individuals. However, his "simplified" results are associated with the consideration of particular regions of that space. More specifically, the search for optimum points may take place by initially considering a small number of utility functions and then moving on to consider positive linear combinations of those functions (see above). Similarly, his propositions concerning the optimality of maximum points or the maximal property of optimum points are illustrated by considering what appear to be special cases (see, in particular, his argument concerning the continuity of maximum points as positive linear combinations of utility functions are varied). If one follows de Finetti's argument, standard propositions of optimum theory appear to be special cases of a much larger possibility set. For example, individuals may search the optimum space by considering different subsets of m_1, m_2, ..., m_k 'initial' utility functions (m_1, m_2, ..., $m_k < n$), or they may decide to expand their search set beyond the domain of positive linear combinations of 'initial' utility functions, or they may want to compare sets of weights ρ_1, ρ_2, ..., ρ_h far apart from each other (thus giving up the continuity property of maximal values). This abstract topology allows de Finetti to outline a general theory of optimal states, and to identify the additional, special assumptions that are normally introduced in the standard economic analysis of those states. In particular, de Finetti suggests that the subjective knowledge of the relevant set of optimum points may be highly contingent, even if local search is justified by the abstract topology of the optimum space (see above). Individuals whose knowledge of optimum states is circumscribed are justified in considering positive linear combinations of "salient" states in as much as optimality properties for a subset of optimum states also hold for the complete set of those states. Since the absolute maximum of any given positive linear combination of utility functions is also an optimum point (see above), individuals with limited knowledge and reasoning ability may concentrate upon a circumscribed set of utility functions, and are justified in "projecting" their direct knowledge of those functions onto the indirect knowledge of positive linear combinations of the same functions. In short, a particular topological structure is found to justify the class of decisions guided by maximization of a positive linear combination of "partial" objective functions. However, what justifies maximization as a means to identify optimum states also circumscribes its use to a relatively narrow set of conditions. For example, a set of n individuals may be associated with optimum states of more than $n-1$ dimensions. In this case, the identification of optima might require the consideration of the objective functions of k additional individuals or social bodies, so that optimum points would generate a variety of $n-1 + k$ dimension. Symmetrically, the objective function of certain individuals might be split into s partial functions (as it may happen with complex, or internally differentiated, identities), in which case the optimum points would generate a variety of $n-1 + s$ dimension. In either case, the identification of optimum states would be significantly more difficult. For example, individuals may be unable to find the absolute maximum of a positive linear combination of partial objective functions that are partly beyond their direct control (case a), or they may be unable to construct a meaningful 'aggregate' objective function from partial functions associated with conflicting objectives they have fully internalized (case b).

The topological structure of the space of optimum positions, and the analysis of rational choice within that space, suggests to de Finetti the distinction between "geometry of utility" and "mechanics of utility":

> The foregoing considerations and hints belong to that part of pure economics we may call economic geometry or geometry of utility, that is, to that part which deals with different economic situations, independently of which economic acts are required to bring such situations about. Pure economics of the classical type, on the other hand, has often being defined as "the mechanics of utility", and its purpose is to anticipate what is going due to the play of "economic forces". Pure economics deals with that problem, as it is well known, maintaining that the free play of such forces should allow the attainment of an optimum position. (de Finetti 1936: 321)

De Finetti's conceptual distinction suggests that there is, in his view, an important difference between the topological structure of the optimum space and the functions describing the seemingly rational behaviour of individuals within that space. As we shall see below, that distinction is central to de Finetti's theory of rational choice[1].

3 Rationality and its bounds: optima, maxima and multiple objectives

De Finetti's proposition that, under particular conditions, there is a one to one relationship between optimum points and maximal states (see above) is the starting point of a complex argument leading to the conclusion that, in many important cases, individual maximization would lead to a solution significantly distant from an optimum point. This is how de Finetti identifies the core issues: "what is relevant is to properly understand the relationship between optimum points and decision problems. To reach a decision we need to choose a maximum point, and we know that such a point needs to be an optimum point ... [I]if what is essential is to maximize the final (objective function) $f(P)$, this is done by choosing $P = P^*$. We know that P^* is but one of the optimum points, and it is not true that $f(P)$ is close to its maximum on the whole variety of optimum points;

---

[1] De Finetti's analysis of optimum decisions with multiple objectives is one of the earliest formulations of the theory of multiobjective optimization (see Rossignoli 1999: 47-53; see also Zaccagnini 1946). For recent treatments of decision theory with multiple objectives, see Keeney and Raiffa 1976; Sawaragi, Nakayama and Tanino 1985; Collette and Siarry 2003.

it is in fact much closer to it on non-optimal points close to P^*" (de Finetti 1975: 646). If we assume that the situation set P is a linear space and that, for every optimum point, there is a corresponding tangent hyperplane, one has to conclude that "$df(P)$ is equal to 0 for displacements dP on the variety of optimum points" (de Finetti 1975: 646). However, "this conclusion is only locally true, that is, in a neighbourhood of that particular point" (de Finetti 1975: 647). A remarkable implication of this property is that the equalization of marginal utilities associated with condition $df(P) = 0$ only holds in the neighbourhood of each particular point. This means that marginal utilities only allow the comparison of any given optimum with other points in its own neighbourhood. Marginal criteria are useless once we want to compare the given optimum with alternatives situations sufficiently distant from it. If each partial objective function $f_h(P)$ entering the maximization exercise is assigned a "price" determined by the condition $df(P) = 0$, the consideration of the manifold "prices" associated with the different partial objective functions allows us to say that the over-all function $f(P)$ is maximized in a particular situation only as long as we are comparing it with alternative situations sufficiently close to it. The above criterion for maximal choice is not a general one, since, as de Finetti points out, "to rely on comparisons in terms of value (rather than in terms of utility) is like measuring the space we have travelled by multiplying initial speed by time (which is obviously an acceptable approximation for short durations only, in which large changes of speed may be excluded)" (de Finetti 1975: *ibidem*).

The above conceptual framework has remarkable implications for economic analysis. Some implications are discussed by de Finetti himself when he considers decision problems in which the duality between maximal states and optimum states is bound to break down. In particular, de Finetti points out that market prices ("commercial prices") would be inappropriate if one wants to compare situations sufficiently distant from one another. This is because distant comparisons would require the evaluation of information relative to different contexts and often beyond the domain to which marginal comparisons are circumscribed. For example, pricing at unit cost (including fixed costs) might write off a possibility that a purely "marginal" evaluation (excluding fixed costs) could still allow. The duality between maximal and optimum states might break down because "distant comparisons" by means of marginal evaluation are logically inappropriate (see above). In de Finetti's view, this does not mean that marginal analysis is relevant for certain comparisons and nor for others. Rather, one should identify a hierarchy of situations and of corresponding decision problems. As we know, marginal analysis is admissible only in the neighbourhood of any given optimum point (see above). If we consider that "prices" are associated with the marginal values of partial objective functions (or "marginal utilities"), we come to the conclusion that certain prices may be appropriate to compare certain situations but not others. Marginal analysis presupposes the identification of an appropriate context. This means that marginal comparisons can only take place between situations belonging to the same level in the hierarchy of problems. De Finetti points out that inappropriate use of marginal analysis (that is, its use across "too distant" contexts) leads to mistaken comparisons and irrational choices.

De Finetti's identification of context as a necessary prerequisite for the discussion of optimum states leads to the distinction between three different analytical steps: (i) the investigation of optimum conditions in general; (ii) the study of optimum solutions subject to competitive conditions; (iii) the analysis of economic choices resulting from freedom of choice. The comparison between different states of the world (de Finetti's "situations") independently of the economic actions that might lead to them allows de Finetti to introduce a pure theory of optimum states independent of institutional or behavioural constraints. This theory allows him to assess specific institutional devices (such as the price-cost equality of the competitive paradigm) against a general normative benchmark (see above). Finally, the discussion of maximizing behaviour under freedom of choice leads him to analyze situations in which economic agents, by following unrestrained self interest, move along a "mirage line" (to use de Finetti's terminology) towards an "attraction point" (also a term used by de Finetti) that is far removed from any optimum state.

It is interesting to see how de Finetti examines the multi-agent setting in which unilateral self-interested choices may lead all actors to attain increasingly worse levels of attainment of their respective objective functions:

> Let us assume that the decisions of any side ... be unilateral...; in this case A can only decide to move along "horizontal lines" ..., which are its "lines of freedom" (the vertical lines, on the other hand, would be the lines of freedom for B); on such lines the most advantageous point for A is the tangency point relative to the indifference curves, that is, the point located on the line drawn from below through V_A [see Figure 3]. In [Figure 4] we see the optimum

line (the segment from V_A to V_B), the two mirage lines crossing one another at point H, which we may call "attraction point", as this is the point at which a sequence of unilateral moves may stop (this sequence would bring further and further away from any possible optimum: from P_o to P_1, as a result of A's move, from here to P_2, by a unilateral decision of B, and then alternately to P_3, P_4, and so on, until point H is asymptotically reached). (de Finetti 1969: 62-63)

At point H, the process described above comes to an end as "it is only at this point that the two agents under consideration have damaged one another so much that they can do it no more with the illusion of an improvement" (de Finetti 1969: 63). De Finetti is aware that the situation described above "is, essentially, the classical duopoly problem, in a different context and as a general scheme" (de Finetti 1969: *ibidem*). Indeed, he refers to the analysis of that problem within the framework of game theory (de Finetti 1969: *ibidem*). However, he also points out that game theory throws light upon this impossibility to attain a solution ("insolubilità") rather than suggesting a way to overcome it (de Finetti 1969: *ibidem*).

De Finetti's discussion of optimum theory emphasizes the variety of optimum states and the hierarchical structure of the space of optimum positions. In particular, he highlights the formal analogy between the maximization of a variety of partial objective functions for any given individual and the simultaneous maximization of objective functions for different individuals. In both cases, it is necessary "to first introduce different functions $f_1(P), f_2(P), ..., f_n(P)$. The final function $f(P)$ will be introduced only subsequently, and after considering the pros and cons relative to any given partial objective. Such a function will obviously be an increasing function of all functions $f_h(P)$. Thus it will be $f(P) = u = \varphi\,(u_1, u_2, ...,u_n)$, with $u_h = f_h(P)$, (with φ an increasing function)" (de Finetti 1975: 645). As we have already seen, de Finetti maintains it is natural to assume that, if we fix the values $u_h = f_h(P)$ for n-1 partial objective functions, "the residual one will have a single point of (absolute) maximum" (de Finetti 1975: 646). However, he also notes that the reciprocal proposition is not always true. This is because "what is relevant here is to maximize the final $f(P)$, and this happens if we choose $P = P^*$. We know that P^* is one optimum point, but it is not true that $f(P)$ is closer to the maximum on the whole variety of optimum points; it is in fact much closer to the maximum at non optimum points in the neighbourhood of P^*" (de

Finetti: *ibidem*). This statement is central to de Finetti's theory of normative structures. Partial objective functions quite naturally lead to the identification of a single maximal state for any given objective function $f_n(P)$ once the values of all other n-1 partial objective functions are taken as given. However, the topology of the optimum space is such that the distance from that particular maximum of the final (or aggregate) objective function $f(P)$ may be greater at other optimum points than at sub-optimal points close to P^*. In other words, once a particular optimum is selected, the maximal state of function $f(P)$ associated with the assumption $f_1(P) =$ constant, $f_2(P) =$ constant,…, $f_{n-1}(P) =$ constant, is immediately determined. However, the value of the same function may be closer to $f(P^*)$ at sub-optimal points sufficiently close to P^* than at optimal states derived by choosing a different constellation of fixed partial objectives. The interpretation of this property is straightforward. In de Finetti's words, "[i]f it is true that the 'optimal' point, relative to any criterion compatible with the 'partial objectives', is always one of the 'optimum' points, it is not true that any 'optimum' point is (for any such criterion) a good or acceptable one. Points close to the best are 'good', or 'acceptable' ones, even if they are not optimum points, whereas optimum points that are further away are not such, and may even be the worse" (de Finetti 1973a: 31-32). This argument is illustrated by considering the "limit case" in which "one of the objective functions is maximized without having regard for all the other objectives (which, as a rule, would be disastrously overlooked)" (de Finetti 1973a: 32). To conclude, in de Finetti's framework optimum theory is primarily an analytical tool suitable to the investigation of situations in which multiple objectives are to be achieved. Optimum theory highlights where choices among conflicting values (or interests) have to be made and where, on the other hand, one is simply looking for the maximal value of a particular objective while taking as conventionally given (and thus normally disregarding) the achievement levels of all other objectives. De Finetti examines the topology of multiple objectives and highlights that, in many cases, attainment levels are to be critically assessed in terms of the underlying value structure. Optimum theory highlights the hierarchy implicit in the maximization of a positive linear combination of partial objective functions. Maximal states for objective n are relative to a particular selection of attainment levels for the remaining n-1 objectives. This property, according to de Finetti, turns optimum theory into a powerful tool for the assessment of normative structures, and more generally for the critical evaluation of particular institutions and socio-economic arrangements[2].

4 Optimum theory and competitive analysis

Since his early economic writings, de Finetti insisted upon the role of pure economics in the critical discussion of value systems and associated policy objectives. In particular, he emphasized the need to undertake a "balanced investigation of the conditions necessary to bring about certain desired economic situations. And this problem, in its full generality that goes beyond the realization criteria associated with different types of economic organization, can be formulated mathematically and should be formulated in that particular way. It is with regard to this specific task and standpoint that pure economics is not sufficiently pure, abstract and simple: it is not sufficiently abstract to move beyond the specific features of a determinate and preconceived economic system, not sufficiently simple to identify the problem in its fundamental essence" (de Finetti 1936: 317; my emphasis). To achieve this objective, one has to expand the core of pure optimum theory by "maintaining it free from the influence of concepts drawn from practice that since its early development give it a different nature and inadvertently lead one to assume what is subsequently claimed to be a result of the analysis" (de Finetti: *ibidem*). In particular, de Finetti maintains that the pure theory of optimum structures allows comparisons only between situations associated with the same *economic condition*, which he defines as "a given complex of circumstances making a whole set of 'economic situations' technically possible" (de Finetti 1952: 685). The distinction between "situation" and "condition" is a cornerstone of de Finetti's conceptual framework, and it is worth exploring such a distinction in more detail. An economic condition is determined by "the set of all 'situations' to be obtained from a given situation by altering the allocation among consumers of individual goods and leaving the produced quantities unchanged, ... or also by varying the produced quantities in a way compatible with the existing productive equipment" (de Finetti 1952: 685-6). In view of the above distinction, de Finetti argues that one could directly compare the different commodity baskets available for purchase by a given individual at given

---

[2] The implications of this view for the construction and assessment of social welfare functions is examined in de Finetti and Emanuelli (1967); see, in particular, de Finetti and Emanuelli (1967: 54-59) and Adriani (2006: 114-118).

prices, or the different collections of commodities that can be produced with a given equipment. However, it is not always possible to unambiguously compare situations associated with *different* economic conditions (de Finetti 1952: 706-7)[3].

De Finetti discusses an important corollary of the above distinction when considering the relationship between equilibrium prices and optimum states. His argument is best illustrated by considering the standard allocation process associated with the so-called "Edgeworth box" (see Figure 3; de Finetti 1973a: 38-41).

In his discussion of the Figure 3, de Finetti calls attention to three distinct propositions:

(i) "for any optimum point P, the common direction of two indifference lines (and thus of their tangent) identifies the price (ratio of exchange), which is unambiguously determined in a neighbourhood of P" (de Finetti 1973a: 39);

(ii) As we move point P along the contract curve , "the corresponding straight line ... varies in a continuous way, so that, as it goes through P_1 and P_2 moving from two opposite sides relative to P_0, it should also go through P_0 moving from a certain point on arc P_1P_2" (de Finetti 1973a: 41);

(iii) "for any original position P_0 ... there are manifold prices that take in a single step to an optimum point (obviously, these optimum points would be different from one another)" (de Finetti: *ibidem*).

In short, it is possible to find an optimum point that can be reached from a particular position by exchanging goods at a specific set of equilibrium prices (proposition ii), and there are in general different optimum points that can be reached from any given position by exchanging goods at equilibrium prices (proposition iii).

---

[3] De Finetti's distinction between "situations" and "conditions" may be related with his over-all conception that "[a] ... viable notion of objectivity lies with a 'deep analysis of problems', aimed at avoiding hasty judgements, superficial intuitions and careless conclusions, to form evaluations which are the best one can attain in the light of the available information" (Galavotti 2001: 173).

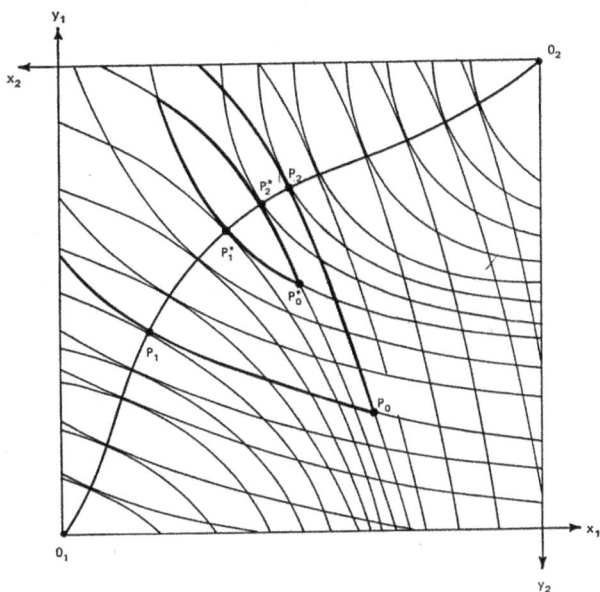

Figure 3. *Optimum points and equilibrium prices (the Edgeworth box)*[4]

The overall endowments of goods x and y (basis and height of the box) may be allocated between agents *1* and *2* according to infinite combinations corresponding to points in the rectangle. Agent *1*'s indifference curves are concave relative to O_1, agent *2*'s indifference curves are concave relative to O_2. Points on the "contract curve" O_1O_2 are optimum points. Points on P_1P_2 (or $P_1{}^*P_2{}^*$) are preferred by both agents to original position P_o (respectively, $P_o{}^*$).

It is generally impossible to find a one-to-one relationship between optimum states and original positions. In general, any given position (such as P_0) is compatible with the attainment of manifold optimum states (through different sets of equilibrium prices), and any given optimum state may be attained from manifold positions. This implies that "prices as ratios between marginal utilities … are meaningful in the neighbourhood of a well-defined point of Pareto optimum" (de Finetti 1973a: 41). For they will be able to remove small displacement form the optimum state through small transactions between agents. However, different sets of equilibrium prices are compatible with different optimum points (starting from the same position). In de Finetti's view, this points to a serious

---

[4] Figure reproduced from de Finetti (1973a: 40).

drawback affecting the utilization of equilibrium prices in the allocation of resources[5]. A visual representation of this case is given in Figure 4:

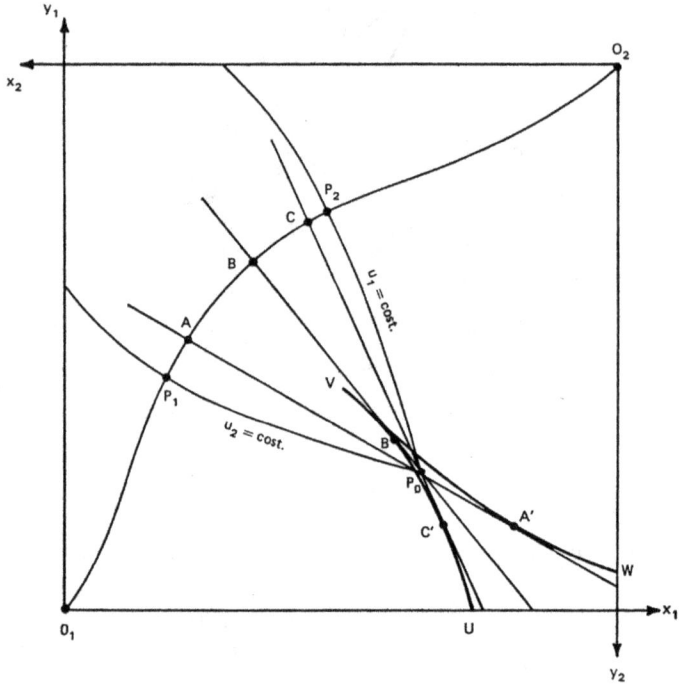

Figure 4. *The multiplicity of "optimum prices" in the Edgeworth box*[6]

Multiple price lines are associated with multiple optimum allocations (such as A, B, C) starting from the same original position P_o.

5 Preference structures and feasible design

In a late writing on the scope and status of economic analysis (de Finetti 1973a, 1973b), Bruno de Finetti emphasized the normative character of economics and called attention to the relationship between the setting of economic and social objectives beyond the domain of existing institu-

---

[5] De Finetti's awareness of this property is at the root of his guarded attitude to the welfare (and policy) implications of competitive equilibrium (see also Lunghini 2007).
[6] Figure reproduced from de Finetti (1973a: 42).

tional arrangements and the feasibility constraints associated with those arrangements:

> It is expedient to state the necessity and purpose of establishing economic science upon utopian thinking ... Utopian thinking in economic science consists ... in examining the possibility of the actual functioning of systems conceived as mental "utopian" schemes. (de Finetti 1973a: 15)

More specifically, de Finetti writes that:

> The specific task of economic science in the utopian formulation that one should promote consists ... in the following: to translate in precise form desired objectives that have been initially expressed in a more or less vague and indeterminate way, to investigate their internal consistency, and to suggest, if that is the case, how to modify or complement them; and to outline forms of social organization aimed at achieving the desired situations ... (de Finetti: *ibidem*)

De Finetti maintains that economic analysis ends up following one or the other of the three following approaches (or "limit-types" in his own terminology):

> *a.* [To assume that economics, both as a science and as a set of phenomena] is subject to immutable concepts and laws; *b.* [To assume that] it is subject to concepts and laws that deterministically depend on historical circumstances that human will cannot master; *c.* [To assume that] it is open to all technically consistent solutions, that is, to all solutions it is possible to conceive. (de Finetti 1973a: 18)

The third point of view entails that economic analysis should set itself the goal of formulating "utopian" set ups, in the sense of "models whose realization one should approximate" (de Finetti 1973a: 18). If this is the case:

> "[V]alue judgements" are the first necessary prerequisite. One has to start off from preference criteria derived from them, ... to take into account objective limiting circumstances (such as resource availability and technological possibilities), and to determine the optimal solution that it is theoretically possible to attain in the absence of institutional constraints. After that, we have to identify the institutional set up, that is, the set of conventions of the organiza-

> tional, legal and institutional type that one has to choose not from a priori or customary ways of thinking but paying attention to the fact that, under the existing circumstances, they would be the most suitable ones in realizing in the most practical way a situation close to the optimal one. (de Finetti 1973a: 19)[7]

One important implication of the above point of view is the intertwining of subjective evaluation and objective judgement under any given set of circumstances. As de Finetti acknowledges, "obviously, any value judgement is a subjective one" (de Finetti: *ibidem*). However, there is for any individual, and more particularly for any scholar, a "commitment to objectivity" (de Finetti), which "consists in correctly drawing conclusions according to declared value judgements, be they shared by himself (herself), assigned to other people, or hypothetical ones" (de Finetti: *ibidem*).

A characteristic feature of normative economic analysis (in de Finetti's sense) is to identify a goal consisting in the maximization of a "final objective function $f(P)$ that shows the degree of desirability associated with situation P. [Such a function], as we have seen, is aimed at collecting in a synthetic judgement the attainment levels of all partial objectives one wishes to consider, as expressed by functions $f_h (P)$ (as already stated, $f(P)$ will be an increasing function of functions $u_h = f_h (P)$, $h = 1,2,..., r$)" (de Finetti 1973a: 34). The partial objective functions considered in this normative exercise are identified as follows:

> The r partial objective functions will be divided into two groups: the n functions expressing "individual utilities" relative to any one of the n individuals of the social group under consideration, and the remaining $r-n$ [functions] that would represent collective needs. The first n functions are (to a certain extent) the same considered in

---

[7] The "utopian" standpoint emphasized by de Finetti in his late writings is in fact a development of a conception that was clearly expressed already in his early contributions to economics. There we find the idea that, in order to make economic analysis an effective instrument for the critical assessment of economic institutions, it is necessary to identify a "core of theoretical principles, invariant with respect to all unnecessary assumptions" (de Finetti 1935: 229; see also Rossignoli 1999: 39-47). A similar view has recently been expressed by Luigi Pasinetti who has proposed to this purpose a specific methodological device called "separation theorem": "[t]his theorem states that we must make it possible to disengage those investigations that concern the foundational bases of economic relations – to be detected at a strictly essential level of basic economic analysis – from those investigations that must be carried out at the level of the actual economic institutions, which at any time any economic system is landed with, or has chosen to adopt, or is trying to achieve" (Pasinetti 2007: 275).

standard welfare economics, in which $f_h(P)$ denotes the degree of satisfaction that the individual assigns to directly experienced benefits. ... There is much indeterminacy as to what one means by P and $f_h(P)$. Any point (situation) P , for the social group as whole, may be seen as an "instant" situation (or almost so) , (that is, a situation associated with) the quantities of goods and services ... available to any one of the n individuals in the unit of time over a period in the neighbourhood of the instant under consideration, or else over a complete time horizon up to a more or less distant future ... The optimum problem, if we consider it by limiting the analysis to individual objectives, corresponds to Pareto's initial approach ... and to the variants in which that approach is presented in the more or less modified versions of Welfare Economics. ... One may identify a variety of social objectives (in the wide sense of "non individual objectives") ... Sometimes the distinction between "individual" and "social" is only matter of convention ... In all cases ... from the mathematical point of view, one has to introduce a new function $f_k(P)$ (for $k = n+1, n+2, ..., r$) ... for any new objective. Nothing changes except that such functions are r instead of n (with the addition of r-n social objectives). (de Finetti 1973a: 34-49)

It is interesting that, in de Finetti's formulation, it is possible, in principle, "to treat [the social objective functions] f_h exactly as all the others in order to determine the points of Pareto optimum" (de Finetti 1973a: 49). However, it is often assumed that social goals are constraints rather than independent purposes to be considered in the same way as individual objectives. In this case, the economist (or the policy maker) establishes that any given "$f(k)$ has to attain a fixed level ($f(k) = u_k$). In this case, one would have a problem of constrained Pareto optimum (that is, an optimum problem for functions f_h, $h=1,2,...,n$, constrained to the region in which functions f_k, $k = n+1, n+2, ..., r$ take on the assigned values.)" (de Finetti: *ibidem*). De Finetti is critical of this approach, which he clearly considered at variance with the basic methodology of optimum analysis. In his view, there is no reason to think that a certain subset of objective functions should be taken as given a priori, and as a consequence excluded from the joint maximization exercise of all individual and social objectives. In this case too, according to de Finetti, it would be important to make use of marginal comparisons: "there will be a marginal utility of variations of u_k to be compared with [the marginal utility associated with] those of u_h , as it is done when all functions are included in a single optimum problem" (de Finetti: *ibidem*). In this way, optimum structures appear to be entirely open-ended and conducive to a rational discussion of goals in view of their mutual dependence and inherent variability.

6   Value systems and rationality constraints

Bruno de Finetti explicitly notes the role of "subjective" value judgements in addressing the specific set of maximization exercises associated with a variety of different objective functions. It is interesting to follow his line of reasoning in some detail. As to the mathematical nature of the problem, he writes:

> One has an optimum problem when we attempt to make two or more functions as large as possible: we thus define optimum points characterized by the fact that it is impossible to move away from them by improving (that is, by increasing) at the same time all the (two or more) functions under consideration. The definition itself states that any optimum point for given functions is also a point of constrained maximum for any one of them subject to the constraint of maintaining constant the value of all the other functions. This implies that the two problems may be dealt with in a single treatment. (de Finetti 2005: 539)

This point of view makes clear the relationship between optimum problems and problems of constrained maximization (or minimization), but also the conceptual distinction between them. Constrained maximization leads to the search of maximal values within a particular region of the relevant domain space. Optimum problems leads to the search of maximal values for one partial objective function while assuming fixed attainment levels for all other partial objectives. In the former case (constrained maximization) the focus is upon "objective" constraints (for example, resource endowments); in the latter case, the focus is upon "subjective" constraints, such as those associated with the existence of multiple goals, or spheres of interest, which are mutually constraining one another. Once a maximization exercise is formulated by making explicit the "internal structure" of the objective function, the problem itself "takes a more differentiated character, and leads to a more constructive point of view" (de Finetti 1975: 645). For example, it is possible to consider the internal preference structure of the same individual, who may set himself (herself) different and sometimes partially conflicting goals. In this case, we may discover that different individuals have the same collection of partial objectives but adopt different distributions of weights across them. Or, we may consider the internal preference structure of a given community, and discover that a collective objective is associated with a two-stage procedure by which social preference presupposes a set of individual objective functions $f_1(P)$, ..., $f_n(P)$, and that such functions "could in turn be generated as a synthesis of other [functions] $f_{ih}(P)$, which ... would denote the preference of individual i for [situation] P relative to partial objective h" (de Finetti: *ibidem*). Last but not least, partial objectives may be interpreted as the same objective under different conditions. For example, a given individual may express different preferences for the same decision P depending upon her (his) own evaluation of the consequences of such a decision in situation h (see de Finetti 1975: 646). The latter case calls attention to the role of risk and uncertainty, as each situation h could be associated with a different degree of unpredictability. In de Finetti's own words, this shows that "it is unnecessary to introduce [the notion of probability] into decision theory as an external factor, as it arises in a natural and most meaningful way from within it" (de Finetti 1975: 647)[8]. To sum up, the decomposition of any given objective functions into its "partial" components highlights the fine structure of value systems and makes it possible to assess a maximization exercise in terms of the more circumscribed goals that may be implicit in it. First and foremost, de Finetti's decomposition criterion allows a critical assessment of the weights attached to different partial objectives and suggests interesting, if largely unexplored, ways to explain different motivations across individuals or social groups[9].

---

[8] De Finetti's "derivation" of probability from within the analytical structure of a decision problem suggests a direct linkage with decision contexts and points to a way in which any given context can itself be considered as the result of a particular collection of situations relevant to the decision maker. In this way, probability judgements are inherently case-based, but "cases" themselves derive from the active involvement of actors. (See also, for a different but complementary perspective, Gilboa and Schmeidler 2001).

[9] Donald A. Gillies and Grazia Ietto Gillies call attention to the formal similarity between de Finetti's view that individuals ought to adjust their probability judgements so as to avoid incoherence, and his view that individuals should be prepared to modify and integrate desired objectives in order to reach some form of inter-subjective coherence: "[s]uppose Mr. A and Mr. B have to divide a cake. Mr. A's desired objective might be to take three-quarters of the cake and leave one quarter for Mr. B. Mr. B's desired objective might be to take three-quarters of the cake and leave one quarter for Mr. A. These desires might be rendered consistent by a process of modification and integration, resulting perhaps in an equal division of the cake" (Gillies and Ietto-Gillies 1987: 198). In the former

De Finetti outlines a formal treatment of this class of problems in terms of the search for maximal values of the positive linear combination of a set of partial objective functions such as u_1, u_2, ..., u_n. This may be seen as follows. Let λ_1, λ_2, ..., λ_n be a set of arbitrary positive coefficients. The global function to be maximized is $u = \lambda_1 u_1 + \lambda_2 u_2 + ... + \lambda_n u_n$, where the coefficients λ_1, λ_2, ..., λ_n denote the weights attached to the different partial objectives. The maximum value of this function is also an optimum point (see de Finetti 2005: 541). In particular, any optimum point will be such that the gradient of the u function is zero. This implies grad $(\lambda_1 u_1 + \lambda_2 u_2 + \lambda_n u_n) = 0$, that is, λ_1 grad $u_1 + \lambda_2$ grad $u_2 + ... + \lambda_n$ grad $u_n = 0$. Given a set of positive coefficients λ_1, λ_2, ..., λ_n, and such that $\sum \lambda_i = 1$, there is a single optimum point associated with the above set of coefficients (see de Finetti 2005: 542). This procedure shows that "the search for optimum points" boils down to "the search for ordinary maxima of linear combinations $u = \lambda_1 u_1 + \lambda_2 u_2 + ... + \lambda_n u_n$" (de Finetti 2005: 542), where the coefficients λ_1, λ_2, ..., λ_n are the standard Lagrange multipliers. It is interesting to note that the coefficients corresponding to any given optimum point are also "derivatives that tell how the maximum value varies as the constraints have variation" (Afriat 1987: 113). In other words, the λ coefficients denote what may be called the "marginal value" of any partial objective function, that is, the "marginal contribution" of any partial objective, or "the extra value obtainable", in view of the final objective, when a unit of attainment of that partial objective is added (see Afriat: *ibidem*). At any optimum point, the Lagrange multipliers take a specific set of values. The above interpretation suggests that these values are the weights associated with the different partial objectives and measure what is to be gained – in terms of global attainment – when one or the other of the partial objectives are privileged.

To conclude, de Finetti's theory of optimum spaces suggests a general conceptual structure encompassing rational decisions subject to circumscribed choice. In particular, de Finetti considers the route to circumscription associated with the decomposition of objective functions into a set of weighted partial objectives. In his reconstruction, choice is "bounded" not by cognitive or informational limitations, but by the system of values (weights) that decision makers adopt (in an explicit or implicit way) when pursuing a variety of partial objectives. This point of view suggests a benchmark for the assessment of multi-objective decisions. It also high-

---

as in the latter case, context provides the cue to a rational decision, be it a probability judgement or a judgement concerning the socially acceptable allocation of resources.

lights where the result of a maximization exercise follows the pure logic of choice and where it reflects the constraints of a specific institutional context. Last but not least, de Finetti's reconstruction of rational decision theory highlights the manifold routes open to economic and social policy once the entire range of rational solutions is taken on board. In conclusion, de Finetti outlines the possibility of economic theory as a discipline in which standard economic rationality dealing with the means-end relationship (instrumental rationality) is combined with a rational discussion of objectives and of their mutual balancing as the weights (and values) adopted by any given society change over time.

References

Adriani, R. (2006). "Bruno de Finetti e la geometria del benessere" [Bruno de Finetti and the Geometry of Welfare]. *Il pensiero economico italiano* 14, pp. 101-121.

Afriat, S.N. (1987). "Lagrange Multipliers". In *The New Palgrave. A Dictionary of Economics*, vol. III, ed. by J. Eatwell, M. Milgate and P. Newman, London: Macmillan, pp. 113-116.

Collette, Y. and Siarry, P. (2003). *Multiobjective Optimization: Principles and Case Studies*. Berlin: Springer.

de Finetti, B. (1935). "Vilfredo Pareto di fronte ai suoi critici odierni" [Vilfredo Pareto and his Present Critics]. *Nuovi studi di diritto, economia e politica* 8, pp. 225-244.

de Finetti, B. (1936). "Compiti e problemi dell'economia pura" [Tasks and Problems of Pure Economics]. *Giornale dell'Istituto Italiano degli Attuari* 7 (3), pp. 316-326.

de Finetti, B. (1937a). "Problemi di 'optimum'" [Optimum Problems]. *Giornale dell'Istituto Italiano degli Attuari* 8, pp. 48-67.

de Finetti, B. (1937b). "Problemi di 'optimum' vincolato" [Problems of Constrained "Optimum"]. *Giornale dell'Istituto Italiano degli Attuari* 8, pp. 112-126. (English translation as "Problems of Constrained 'Optimum' " in *Italian Economic Papers* III, ed. by L. Pasinetti. Bologna and Oxford: Il Mulino and Oxford University Press, 1998, pp. 261-277).

de Finetti, B. (1952). "Sulla preferibilità" [On Preferibility]. *Giornale degli economisti e annali di economia* 11, pp. 685-709.

de Finetti, B. (1969). *Un matematico e l'economia* [A Mathematician and Economics]. Milano: Franco Angeli.

de Finetti, B. (1973a). "L'utopia come presupposto necessario per ogni impostazione significativa della scienza economica" [Utopian Thinking as a Necessary Prerequisite for any Meaningful Approach to Economics]. In *Requisiti per un sistema economico accettabile in relazione alle esigenze della collet- tività*, ed. by B. de Finetti. Milano: Franco Angeli, pp. 13-76.

de Finetti, B. (1973b). "Utopia, as a Necessary Presupposition for Every Signifi- cant Foundation of Economics". *Theory and Decision* 5, pp. 335-342.

de Finetti, B. (1975). "Due lezioni su 'Teoria delle decisioni'" [Two lectures on Decision Theory]. In *Contributi del Centro Linceo interdisciplinare di scien- ze matematiche e loro applicazioni*, n. 6, Seminari su la scienza dei sistemi, Part Two, pp. 643-656.

de Finetti, B. (2005). *Matematica logico intuitiva* [Logical intuitionist Mathema- tics]. Milano: Giuffrè (1st edition: 1943; 2nd edition: 1956).

de Finetti, B. and Emanuelli, F. (1967). *Economia delle assicurazioni* [Econo- mics of Insurance]. Torino: UTET.

Galavotti, M.C. (2001). "Subjectivism, Objectivism and Objectivity in Bruno de Finetti's Bayesianism". In *Foundations of Bayesianism*, ed. by D. Corfield and J. Williamson. Dordrecht-Boston-London: Kluwer, pp. 161-174.

Gilboa, I. and Schmeidler, D. (2001). *A Theory of Case-Based Decisions*. Cam- bridge: Cambridge University Press.

Gillies, D.A. and Ietto-Gillies, G. (1987). "Probability and Economics in the Works of Bruno de Finetti". *Economia internazionale* 40, pp. 192-209.

Keeney, R.L. and Raiffa, H. (1976). *Decisions with Multiple Objectives*. New York: Wiley.

Lunghini, G. (2007). "Bruno de Finetti and Economic Theory". *Economia poli- tica* 24, pp. 3-11.

Pasinetti, L.L. (2007). *Keynes and the Cambridge Keynesians. A 'Revolution in Economics' to be Accomplished*. Cambridge: Cambridge University Press.

Rossignoli, C. (1999). "La schiavitù dell'anarchia. Gli scritti di Bruno de Finetti sull'equilibrio economico" [The Slavery of Anarchy. Bruno de Finetti's Writings on Economic Equilibrium]. *Economia politica* 16, pp. 35-64.

Sawaragi, Y., Nakayama, H. and Tanino, T. (1985). *Theory of Multiobjective Optimization*. Orlando: Academic Press.

Zaccagnini E. (1946). "Massimi simultanei in economia pura" [Simultaneous Maxima in Pure Economics]. *Giornale degli Economisti* N.S. 6, pp. 258-292.

9

Why de Finetti's Critique of Economics is Today more Relevant than Ever

GRAZIA IETTO-GILLIES [*]

1 Introduction

1.1 A personal recollection

I am very pleased to have been invited to prepare this paper because it has given me the opportunity and incentive to revisit some of the works by my old teacher and to write about him, his contribution to economics and his vision for the economy and society.

De Finetti played an important part in my life. I studied with him a few years after he became Professore Ordinario at La Facoltà di Econo- mia e Commercio, La Sapienza University in Rome. There were three courses he was responsible for: Matematica Generale, Matematica Finan- ziaria and Matematica Attuariale in the first, second and third year of our studies; he lectured on the first and last course. He also supervised my Tesi di Laurea on decision theory in inventories and the Poisson distribu- tion; this meant that I was in contact with him for quite a while. More- over, after my graduation, I attended the Summer Schools he organized in L'Aquila and Frascati; at the former one we had Ragnar Frisch, Malin- vaud and Morishima among the speakers, while at the latter one I met Savage.

[*] I benefited from the lively discussion following the presentation at the International Workshop in Bologna. I would also like to thank Nicola Acocella, Donald Gillies, Roberto Scazzieri and Mario Tiberi for useful comments on earlier drafts of the paper.

Bruno de Finetti, Radical Probabilist
Maria Carla Galavotti (ed.)

Most first year students found the material and his delivery difficult to follow and indeed they were. His soft voice was often inaudible in the large, crowded, old theatre of Piazza Borghese where the Faculty was located at the time; like many mathematicians, he often spoke while writing formulae on the blackboard: no OHP or PP presentations then! The material presented in the first year was not easy for many of the students with a poor mathematics background.

However, some of us felt from the very beginning that there was something special in him and his course; it was partly how he made the content relevant with the aid of examples and partly his presentation of the material; he was always trying to lead us to the essential points in the arguments and gloss over the more technical and specific ones. I particularly liked his blending of formulae and graphs; analysis and geometry: we were always asked to try and represent graphically the algebraic formulations. Some of us began arriving early for his lectures to make sure we could secure a seat in the first few rows.

By the third year most students had dropped off the course – as was then the norm in Italian Universities and to a large extent still is – and there were, blissfully, few of us following his unit on probability. This was the time when he did with us his experiment on probabilities assessment: we were asked to assign probabilities to our predictions for the forthcoming football matches. Following the Sunday football matches, he would discuss with us the actual results against our own assessments. The porter in charge of our classroom always scored best: much better than us students and the probability Professor. De Finetti told us that for this experiment he was liaising with a Professor Lindley of Britain.

I also remember an amusing anecdote he told us at the start of his probability course when he warned us not to mix up probability with either superstition or religion. His story relates to an episode that occurred in a Church in the north of Italy; I do not remember whether he or one of his relatives or family friends was in the congregation. The priest was giving a sermon against the combined evils of superstition and gambling; he was against both. Moreover, he was trying to impress on the congregation the irrationality of superstition[1] in relation to gambling on the "lotto" by saying that it was pure superstition to believe that if he mentioned

---

[1] Many Italians believed that dreams could be interpreted in terms of numbers which would come out in the weekly "lotto" draw. There were experts and indeed books on the art of numerical interpretation of dreams. For example, it was common belief that if a dead person spoke to you in your dream, you should play 47.

three numbers they should come out in the lotto draw. He did mention three numbers at random; a member of the congregation did play them immediately after and they did come up. Both de Finetti and his students roared with laughter and bonds between them were further cemented.

By then he had become much less daunting for me and my sister, who was studying with me at the time, also for a specific personal reason. Our home in Piazza Annibaliano was along the same bus route as his in Via Poggio Catino: the 58 bus leading from Piazza San Silvestro near the Faculty where he taught and we attended courses. We would often meet him on the bus and he was always very courteous towards the two shy sisters.

However, de Finetti had also a dramatic effect on my private life. It was one evening in February 1970 in Cambridge when at a small party in my room, one of the guests brought along a friend of his, a Donald Gillies. We both tried small academic talk and thus I learned that he had just finished his doctoral thesis on Philosophy of Probability. I asked him whether he might have come across the name of my Professor in Rome, a Bruno de Finetti. He almost jumped up from the floor where we were all sitting. It gave us an excuse for meeting again, and again. After 36 years we still, occasionally, discuss de Finetti and his work.

1.2   Reading de Finetti on economics then and now

Apart from some remarks here and there on decision theory during the course on probability theory and during my work for the thesis, I did not hear much from de Finetti on his views on economics. My first proper encounter with them was during the summer schools which were specifically on economics-related topics.

I have to confess that at the time – the late 1960s – I felt that his views were out of sink with the prevailing views that I was by now becoming familiar with in my further studies of economics, the subject I had, by then, switched to. Macroeconomics was the order of the day while he was putting forward a micro, Paretian type economics. Moreover, he was criticising the neo-classical economic theory; but that seemed to me to have been so discredited that it was, perhaps, not worth considering. At the time most progressive people in economics were convinced that the neo-classical system was wrong and that it was gradually being recognized as such.

Over the last year I have read or re-read de Finetti's economics works and my reaction now is that they are very relevant today; indeed more relevant than when he wrote them: it is as if the economies and economics have moved much more along the lines he had been criticizing. The next sections will explain why I have now moved to this position.

2 De Finetti's writings on economics[2]

Why was de Finetti interested in economics? The question is relevant because the answer is connected with understanding his approach to economics. In his (1969: Ch. 1) he gives two main reasons: the first is a desire to attack prevailing fetishes coming from "...his moral revolt against acquiescence to distortions" (1969: 26) found in society and economies. Put it slightly differently, this is the desire to:

> ... create an economic system that is efficient and aims at meeting the needs of society in an environment of social justice and free from those distortions that derive from sectoral pressures, partisan interests, egoism of privileged people, internal contradictions of the private structure on which the economy is based. (de Finetti 1969: 29)

From this dissatisfaction derive his reasons for studying economics; they are contained in various points in his writings and particularly in the formulation of the following question:

> How could we modify the rules of the game so that the apparent antithesis between the interests of each individual considered in isolation and those of each individual seen as member of society disappears? (de Finetti 1969: 90)

Moreover, he sees economics as a study of problems and institutions in order to change institutions, priorities and outcomes (de Finetti 1969: 87). Essentially his study of economics comes out of a desire:

- For social justice.
- For efficient systems and institutions.
- To use his mathematical skills to show that it is possible to eliminate the contradictions between the interests of individuals as such and those of individuals as members of communities and societies.

---

[2] Most of the comments on de Finetti's works come from his 1969, 1973 and 1976 which are all collections of essays some published previously. To help the reader trace the historical roots of de Finetti's views the appendix gives details of the first publication of the essays. The translation of passages in this paper is my own.

On the latter point it is interesting to note that his use of mathematics in economics parallels his use of it in probability theory. In the former case mathematics is to be used to smooth out contradictions in the preferences of each individual – within her/his own sets of choices – as well as those between individuals. In the case of probability it was through the elimination of contradictions that the Dutch Book could be avoided and thus a system of probability compatible with the basic Kolmogorov axioms could be developed.[3] Not surprisingly, he notices the increasingly close "... points of contact ... between various aspects of economics and applications of probability calculus and statistical methods" (1969: 26).

The major areas of economics considered in his writings are the following:

1. Critique of "classical" theory.
2. Efficiency issues, particularly in the Italian public sector.
3. Policy perspective; social issues, social justice; humanity in dealing with economic issues.
4. Uncertainty/risk/social security issues.
5. Methodological issues in economics.
6. Role of mathematics in science in general and in economics in particular.
7. Positive versus normative economics.
8. How to use mathematics to achieve the aims he wants in economics: operational research; welfare functions and Pareto optimality.

2.1 Critique of the neo-classical theory, efficiency, policies and uncertainty

De Finetti, like Keynes, uses the term "classical" economics to indicate the system of orthodox economics usually known as the neo-classical system/theory. He thinks that this theory is too abstract and does not take account of the historical context; there is too much reliance on ad hoc explanation to fill the gap between theory and reality. Moreover, the neo-classical economists have been unable and unwilling to deal with the conflicts and contradictions between the interest and preferences of individuals and those of society of which the same individuals are members; one of the consequences of this approach is the fact that the conflicts between the present and the future are not given enough weight: the future and the

---

[3] I owe this point to D.A. Gillies.

social are usually sacrificed for the sake of the present and the individual. According to de Finetti the overall result is that in the end society as a whole – or most people within it – suffers.

Two types of conflicts and contradictions keep recurring in de Finetti's writings: (1) external economies and diseconomies; and (2) contradictions between the effects of individual versus collective behaviour.

External economies or diseconomies are those effects of individual behaviour – including contractual agreements – that spill over third parties or society as a whole; these externalities can be positive or negative. The price emerging from the free unhampered working of the market cannot allow for externalities be these positive or negative.[4]

The following are examples of negative externalities.

– The price of the chemical products sold by my firm will take account of the private costs of producing them but not of the social costs – health hazards – of the pollution generated in the rivers, air etc.

– The exhaust fumes from my car pollute the environment causing problems for other people; yet the price of petrol does not reflect this negative externality, neither does the price of cars reflect the diseconomy of noise and traffic.

– If I contract an infectious disease, I can pass it on and society as a whole is in receipt of a negative externality from me: this means that it is in the interest of society as a whole – as well as of myself – that my health should be protected.

There are also many cases of positive externalities:

– If I maintain my garden beautiful, my neighbors and the passers by benefit from it.

– The education of children benefits them and also society including future employers.

– There can be important positive externalities on the production side: the investment in training employees by one company will benefit others – sometimes competitors – when workers move to new jobs.

– Firms learn from the environment in which they operate – from customers, suppliers, distributors; moreover those firms that have production activities in many counties – the transnational compa-

---

[4] For a clear exposition of the theoretical debate on market failures and externalities see Acocella (1998: Part 2).

nies (TNCs) transmit knowledge to a variety of locations via their internal networks: this is an important issue in the contemporary world in which TNCs are becoming ever more relevant in production activities. (Zanfei 2000; Frenz and Ietto-Gillies 2007)

Contradictions between individual and collective behaviour: the unfettered pursuit of individual gains may rebound against everybody. Egoistic behaviour may be legal and above board or it may be the result of cheating; in either case it may backfire. De Finetti gives the following examples. (a) In a theatre hall, a member of the public may be tempted to stand up to get a better view; however, if everybody stands up nobody is in a better position. (b) If an individual employer manages to lower wages and retain his work force the demand for his products may remain unchanged provided the rest of the economy retains higher wages; however, if all employers cut real wages demand will fall for all of them and they may end up worse off. Similarly an employer that may raise wages for her/his employees in the hope of increasing demand for her/his product, can only succeed if other employers raise wages as well.

De Finetti is very critical of neo-classical economics also on the ground that it favours the *status quo*. The so called optimum solution of neo-classical economics are based on the existing distribution of resources; he challenges the assumption that such allocation must be taken as given because economists should not express value judgments on what is the best or acceptable level of income and wealth distribution. Though he accepts Pareto's methodology, he is very critical of the Pareto optimality principle: points of Pareto optimum can be very bad for society. His view is that the incorporation of value judgments is exactly what economics should be about and that mathematics should help to formulate and solve the technical problems deriving from it.[5]

He is also against unbridled capitalism of free market economies on the basis that business decisions led only by the profit motive (de Finetti 1976: 19) encourage the production of useless or damaging products: e.g. cigarettes; ever more fast cars; or to the destruction of agricultural products to support the incomes of farmers in developed countries while people in developing countries starve. To illustrate his dislike of financial and profit motives as the main aims in society and his abhorrence for the inequalities to which they lead, he uses a folk saying from his native Tri-

---

[5] De Finetti's views on his normative Pareto system are clearly highlighted in Adriani (2006) as well as in Chapter 8 by Scazzieri in this volume.

este: "the devil shits on the biggest mound" (de Finetti 1976: 22) and he specifies that money is the devil's shit.

He is also sarcastic on the justifications usually given of the unfettered market mechanism on the basis of freedom. He writes:

> A curious rhetoric may in fact continue to identify as "freedom" any organizational deficiency. ... But, as a human reality – the only one that counts – what is such freedom? ... it is freedom for a privileged minority who have the means needed to take advantage of it ... true and supreme freedom for individuals is that they should all be liberated from the worries and distortions that nowadays poison our existence for opposite, though parallel, reasons those of us who are afflicted by possessions and those afflicted by the lack of possessions ... (de Finetti 1969: 86).

He concludes that free market economies far from being efficient are highly *inefficient* because they produce many useless or damaging products and because they lead to prices that do not account for negative externalities and for the social costs deriving from them. Moreover, there are many in-efficiencies due to poor/perverse institutional structure particularly in the Italian public sector. He thinks that the skills of scientists – including economists and mathematicians – should be used to advise on the best institutional structures for both private and public sectors. A whole chapter (de Finetti 1969: 17) is devoted to attacks on the bureaucracy and the waste of citizens' time it leads to. He proposes a centralized system for personal identification to be used in all dealings between public institutions and individuals, from the collection of taxes to the payments of social benefits: essentially what was later introduced as the Codice Fiscale which already existed in Britain as the Social Security number. Another institution that comes in for a strong attack on the part of de Finetti is the Italian university system for excessive bureaucracy, low respect for the needs and rights of students and misuse of power by some academics.

At several points he considers *policy issues*. He is in favour of government interventions to correct the negative effects of externalities as well as to correct what he sees as grossly unequal distribution of resources. He writes (de Finetti 1976: 19) that "... nobody who is not aware of the real situation on men on the planet Earth (think of a Martian) could possibly advise a distribution as unequal as the existing one and, worse still, a distribution that leaves so much of humanity in dis-human conditions."

He thinks (de Finetti 1973) that most people – or all reasonable and honest people – would agree with the following aims for society:
–   Lower degree of inequalities.
–   Lower degree of social insecurity.
–   Lower degree of reliance on markets and competitiveness.
–   Maximum freedom of choice for individuals.
–   More efficient and less complicated institutions and processes.

He puts his common sense as well as his mathematical skills to the suggestions of solutions. For him economics is and must be "normative" that is it must suggest acceptable and feasible solutions for correcting distorsions and indeed for proposing better institutions and systems. Specifically, he argues along the following lines.

–   He wants a normative version of the Paretian approach. Pareto type of analysis can be applied not to accept the status quo but to show how feasible solutions can be achieved taking account of the preferences of each of us as individuals and as members of the community and society. In the latter role we should all express our preferences; the task of the scientists (economists and mathematicians) is to point out possible inconsistencies and – where necessary – lead us towards more consistent choices.

–   Decision theory and normative economics go together. The techniques of Operational Research can and should be applied to the development of public policies. Operational research can also assist in the development of the individual and socially best system in a variety of fields.

On the latter point he deals with a specific example of great interest to him as a probabilist and social scientist as well as to us all. He devotes a chapter (de Finetti 1969: 16) to Social Security. He discusses the issues of uncertainty and risk in a variety of social areas such as unemployment, health, old age or accidents. What is the best (i.e. most effective and efficient) insurance system – for society as a whole – for each and all of us? What is the best method – for society as a whole – of funding the cost of such insurance(s)? He argues in favour of a unitary system with a single public institution in charge of it in which the large numbers and variety of risks lead to "compensation of risks". Moreover, the best way to fund the cost is via the fiscal system in which people pay on the basis of their ability rather than on the basis of their individual risk. This system would also lead to a redistribution of resources from the less to the more needy people: a move he greatly favours.

Essentially his view of economics can thus be summarized. In line with his approach on probability, de Finetti feels that economics should start from individuals and their preferences and build up to a system in which the choices for society as a whole are consistent with individual preferences. Microeconomics is undoubtedly his preferred method of analysis: for an understanding of the real problems facing people we should resort to the micro economy and analyze the behaviour of individuals with their specific preferences and value judgments. He sees macroeconomics as useful for analyzing broad trends and as a general guide for understanding the system; however, aggregation hides more than it discloses and should not be used as the sole methods for policy planning. Input-output analysis is useful for understanding the linkages between sectors: "With ... precautions, the input-output analysis is certainly an appropriate bridge between macroeconomic generalities and microeconomic perfections" (de Finetti 1969: 217). Nonetheless he does not exclude a priori any of these three methods: we could say that he is in favour of methodological pluralism. He writes (de Finetti 1969: 226): "None of the methods of analysis ... should be excluded: the methodology of macroeconomic models as a preliminary attempt, the method of a preliminary summary organisation in a Leontief type framework, and then that of an effective scrutiny in disaggregated, detailed, localized terms".

The latter can be translated into individual welfare functions which must relate to two sets of preferences: those that we all have as individuals and those that refer to us as members of society, i.e. those preferences that we would like to see in the social system. Neutrality does not exist in real life; it is only a convenient assumption in neo-classical economics. Therefore value judgments about all aspects of society are not only acceptable but must be encouraged, thus the welfare functions will take account of individuals' preferences for society as a whole (for example, individual preferences on income distribution which should not be taken as given and unchangeable). On this point the task for the economist and mathematician is to bring to light possible contradictions and get people to reformulate their preferences on the basis of those contradictions. The pretence of an economic science without value judgments is just that: a pretence done with the purpose of perpetuating the *status quo* and smuggling in the values and interest of the establishment.

2.2   Methodology, mathematics and economics

Scattered in many of de Finetti's writings are many remarks on methodology of science and on the proper use of mathematics in economics. One of the main points he stresses is the role of the scientist in determining aims: the scientists must be involved in value judgments. His 1976 book has the following title: "From utopia to the alternative" and indeed in the very first chapter he argues in favour of a utopian view for the economy and society. He encourages people to move away from a vision of society constrained by what is seen as realistic; this often means a view which is in line with the interests of the existing establishment and the status quo. Instead we should think of the ultimate aims we want for society and then on the means to achieve them. He clarifies that his "Utopia" is not to be taken as "something absurd and unrealizable, rather as a preliminary and idealized description of a future to be constructed as far as possible in line with such a model" (pp. 8-9). He wants people to go into uncharted territory, think the unthinkable if the aims justify it: eventually the unthinkable becomes realizable and realized.

He thinks that science would not have progressed or continue to progress without utopian thinking, without the scientist being able to make a qualitative jump away from prevalent systems, theories and environments be they in the realm of sciences, technology or society. He wrote in 1965 on these issues (de Finetti 1969: 250):

> I thought I might be too optimistic, some ten or fifteen years ago, in foreseeing the development of computers and their applications, in general and in Italy in particular; others probably thought I was mad; and yet, those mad predictions of mine, today might appear to you those of the most retrograde and skeptic madman one can possibly imagine (in the opposite sense).

He meant that, if anything, he had been too conservative in assessing the potential of the computer, not too radical. And later on the same page:

> Reality is the fruit of the fantasy of those people who with fantasy truly build ... and, with their fantasy, they can and do go forward, because gradually fantasy is transformed into the reality of new projects, new principles, new approaches and this process generates new platforms for new constructive leaps of fantasy.

To this writer it seems that when talking about utopia he has in mind something like Kuhnian revolutionary science applied to social issues; he wants to encourage economists to move away from "normal science" approaches and go into "revolutionary science".

He warns against a-historical approaches in which: "... every epoch and every doctrine may present its own conceptions as 'truths' in an absurdly absolute sense without clarifying their historical, relative and provisional values" (p. 33).[6]

He also warns against assumptions that are too unrealistic; some assumptions are needed but they must have the basis in the real world. The relevance of these remarks could be seen in the context of the methodological debate on instrumentalism in economics of which, however, de Finetti may not have been aware. They were sparked off by Friedman (1953)'s claim that realism in economics does not matter; we should accept unrealistic assumptions provided we think that the world behaves *as if* they were true. Indeed for Friedman:

> ... the relevant question to ask about the "assumptions" of a theory is not whether they are descriptively "realistic", for they never are, but whether they are sufficiently good approximations for the purpose in hand (p. 15); ... The decisive test is whether the hypothesis works for the phenomenon it purports to explain. (p. 30)[7]

De Finetti is also skeptical on the use of analogies between economics and the natural sciences be they biology or mechanics. There are grotesque errors on the part of those scientist who "... think of modeling economics or sociology on the theories of mechanics and therefore believe in the possibility of a spontaneous equilibrium in an economic and political regime of anarchic liberalism ..." (de Finetti 1969: 41).

Whether it is appropriate to use mathematics in economics and to what extent, is not something that can be judged in the abstract and a priori: it all depends on what use one makes of it and whether mathematics is useful for the problem in hand. Mathematics can help to reach sensible solutions; however, in order to reach sensible and exact conclusions it is not enough to resort to the use of mathematical techniques (1969: 30). What he likes to see is mathematics as a constructive instrument which confronts real problems rather than pseudo ones; mathematicians must work in symbiosis with economists and not as outsiders: what we need between

---

[6] The historical, relative and provisional conception of truth is affirmed more strongly in his 2006 book discussed in the introduction to this volume.

[7] Among the people who have contributed to the debate are Musgrave (1981) and Samuelson (1963).

the two sets of scientists is collaboration in the sense of full communication, coordination and co-penetration to problems. In a paper inspired by the first World Conference of the Econometrics Society in 1965 in Rome, he writes:

> Any mathematical development, any mathematical critique may emerge from requirements or curiosities that are purely mathematical in nature; however, if they do only this without making us ponder on the economic interpretation they cannot be of relevance to economics; they risk giving rise to confusions by suggesting incautious and gratuitous interpretations. (de Finetti 1969: 176)

De Finetti criticizes the use of mathematics in conventional economics, where often "... mathematics and analysis in particular are seen as all powerful instruments, capable of giving precise and unique answers from complicated developments based on uncertain, vague and approximate hypotheses that are translated in formulae with a large margin of arbitrariness" (p. 64). He also invites the researcher to be aware of the many hypotheses built in into estimation methods (de Finetti 1969: 258).

In (de Finetti 1976: 25) he tells a personal anecdote. "It so happen that on one occasion, while traveling, I found myself in the same compartment with a gentleman who was accompanied by a young man and a young woman with whom he was conversing with airs of a cultured and important person. One of the longest and articulated discourses related to ... astrology of which he maintained the scientific character because it is based on... trigonometry!"

He is very doubtful about the possibility that game theory might solve real economic problems. In a chapter dedicated to this subject (de Finetti 1969: Ch. 7) a section is devoted to "Value of negative lesson" (131-5) he follows and supports Rapoport (1962) in the following elements. (a) Game theory may be useful but not because of the solutions it leads to, rather because it brings to light the different types of reasoning possible in different situations. This leads to more general methodological remarks about the value of negative learning as well as about the relevance of problems raised by theories as well as problems solved by them if any. (b) Game theory brings to light the fact that solutions do not always exist. (c) Decisions based entirely on self interest, lack of trust and cheating can lead to disasters for all parties as in the Tosca-Scarpia case. (d) The most important contribution of studies on game theory is due to the fact that this type of analysis reveals its own limitations.

In his 1973 he puts to use his mathematical skills (algebra and graphs together in his favourite analytical and didactic method) to illustrate how his "utopian" vision for the economy and society can have real solutions; individual welfare functions take account of choices of each of us as individuals and as members of society.

3   Why de Finetti's writings are so relevant today

The last few years of de Finetti's life saw the beginning of significant changes in the economies of the world and in economics. It all began with the Thatcher and Reagan revolution which later spread to other countries with a variety of intensities and speed. It spread – and often was forced on – to developing countries by the so called Washington consensus that is by the consensus of policies of the International Monetary fund (IMF), the World Bank and the US Treasury. It spread to developed countries via the deliberate policies of national governments or the pressure/imposition on them by supranational or international institutions such as the EU or the World Trade Organization (WTO).

I shall discuss the changes in the contemporary economic and social world by dividing them into two parts: (1) changes in the economy and society; and (2) changes in economics. The emphasis is on those changes relevant in terms of de Finetti's view of society and the comments I shall make relate the relationship between his views on specific issues and the changes we have seen or are seeing.

3.1   Changes in the economy and society

The greater integration of the world's economies and societies – the so called globalization process – as well as the new technologies has led to an increased relevance of *externalities*. Some externalities operate at the same time and within generations, some extend into the future and thus affect future generations. Here are some examples of how the economic system in the globalization era has led to increase in a variety of negative externalities.[8]

–   As already mentioned, health and diseases have always been a source of externality: if poor people are more liable to illnesses, there is a danger for them as well as for rich people who may be infected by them. However, there are now stronger and wider *health*

---

[8] On the so called global public goods and ills which lead to financial, environmental and health externalities cf. Acocella *et al.* (2004: Ch. 4).

externalities as a result of stronger integration between the countries of the world: in the post WWII decades tuberculosis was considered to have been eliminated in the developed countries; however, it is now coming back as a result of the movements of people to and from countries where it is still endemic. This is true of other diseases be they old ones such as malaria or relatively new ones such as AIDS.

–   *Environmental externalities.* This is one of the strongest elements of externalities: the high consumption of fuel by some countries has negative effects on us all and indeed on future generations.

–   *Financial crises* spill over more quickly and involve a larger number of countries.

–   *New technologies* have brought their own spillovers: for example noise spillovers from incessant talking on mobile phones.

–   *Social and emotional spillovers.* As communications and television viewing has increased world wide, images of different societies – even if not always accurate – are brought to other parts of the world bringing feelings of unease which lead to desire for better economic, social conditions or democratic and human rights in countries deprived of them. This causes emotional spillovers effects – in viewers sitting far away in front of their television sets – from images which would not have been available decades ago.

De Finetti argued that one of the reasons why government intervention is needed is to correct market failures due to externalities. The need for intervention has now become even more pressing than in his time. However, such intervention is much more difficult now for two reasons. First, because the political environment has become much less interventionist, or rather less willing to correct market distortions since the market is seen as an impartial and efficient mechanism for allocating resources. Second, because externalities that originate in the global sphere require international cooperation and appropriate international institutions fit for the purpose: those institutions that exist were established with different purposes in mind and far from solving the problems they often exacerbate them.

Moreover, de Finetti was critical of the unfettered working of the market; yet we have seen an increase in privatization and *marketization*: goods and services that used to be produced by the public sector are now produced through the market.

As I already mentioned one of the issues on which de Finetti felt passionately strong was the *efficiency* of the overall economic system and of specific institutions, particularly Italian public bodies. Has inefficiency increased? Difficult to say, of course, but my gut feeling is that it has for the following reasons. As western societies have become more affluent, there has developed a struggle by large corporations to induce us to consume more and more products often useless or even harmful ones.[9] A case in point is the food industry. Western consumers are urged to consume food with high calories and fat content: we have death by obesity at the same time as people in many developing countries die of starvation. The obesity pandemic is a relatively new phenomenon – developed largely after de Finetti's life – and it is due mainly to problems in the economic and social system. In echoes of Galbraith's *The Affluent Society*, de Finetti writes on the issue of balance between necessary and superfluous products:

> The private sector, particularly if left in the state of crazy atomism demanded by its zealots, can only give us that wonderful "miracle" that in exchange for lack of necessities, allows us to wallow in excess of the superfluous. (de Finetti 1969: 100)

There is another development of the last two decades that may have increased inefficiencies: the practice of outsourcing or externalization of parts of the production process, usually the non-core part. Some material components or services are outsourced through a variety of contractual arrangements by large private and public institutions to smaller ones. This has led to the fragmentation of production[10] with great difficulties in coordination as well as – in many cases – higher costs (as the British National Health System is experiencing). It is a heaven for lawyers who are called in first to draft contracts and later to resolve conflicts between principal and contractor.

Moreover, there seem to be a rampant culture of control; the new technologies have increased our ability to collect and elaborate data; this is often used to increase control and develop more and more performance indicators and targets which often have the opposite result from the intended ones: these indicators distort behaviour and deflect it away from

---

[9] This statement does not mean to detract from the fact that the new technologies have brought us some new useful products.

[10] The relationship between the organizational and locational fragmentation of production and the labour force is discussed in Ietto-Gillies (2002: ch. 6 and 2005: ch.15).

the objective of doing one's professional job and providing a professional service into meeting targets (see the case of the British NHS or indeed the effects of the Research Assessment Exercise on British universities and on research in general[11]).

De Finetti was very passionate about issues of *social justice* and he had a vision of the world as composed of individuals who lived and behaved as part of a community and society. In his analysis the starting point was individuals; however, their behaviour had to be seen within – and shaped by – the need of the community and society. What developments have there been in the last two decades?

We can certainly say that the Thatcher revolution has greatly enhanced the culture of individualism: one of her famous statements was that "there is no such thing as society".[12] The change in culture seems to have brought with it – among other problems – a decline in the level of trust between people. De Finetti emphasis on trust comes out very clearly in the course of his discussion of the value of Pope John XXIII's *Pacem in Terris*; in that passage he sees the need for people to: "… induce others to listen by listening, gain understanding by showing understanding, …" (de Finetti 1969: 137).

Layard (2005: 81) reports the results of a survey in which people are asked the following question in 1959 and in 1998: "Would you say that most people can be trusted – or would you say that you can't be too careful in dealing with people?" In the earlier date 56% of Britons reply "yes" while only 30% give the positive reply by 1998. For the US the two percentages are 56 in the mid-1960s and 33 in the later period.

On the issue of *insecurity and uncertainty*, de Finetti argues that individuals within society must be made to feel more secure. He wanted an end to temporary university contracts (de Finetti 1969: Ch. 17) and an efficient, fair, secure social security systems that addresses all sources of economic insecurity from unemployment to health to old age to accidents. Well, uncertainty and insecurities have increased tremendously in the last two decades: particularly those related to employment – generated by the so-called flexible labour markets – and to old age.

An example which involves issues of both insecurity and negative externalities is the right of citizens to carry arms in the US. With the gun the

---

[11] Using several examples from the history of science, Gillies (2006) argues that the RAE discourages innovative research of the type that might lead to new Kuhnian paradigms.

[12] Interview in *Woman's own*, 1987.

individual may acquire a feeling of personal security but only at the cost of much greater risks and insecurity for society as a whole.[13]

One of the main differences between the world today and the world in de Finetti's time is the change in income and wealth distribution and the increase in *inequality* within and between countries. De Finetti was appalled at the inequality existing when he was writing; changes in income distribution has certainly not gone in the direction he advocated, on the contrary the rich have become richer and the poor poorer. Indeed there is now a greater reliance on the voluntary and charity sector, something he was strongly against (de Finetti 1973: 46) because, in his view, such reliance puts the needy in a humiliating position.

Let us see how inequality and poverty that so much worried him have changed since he was writing. Economic inequality can be considered between countries and within countries. It is generally agreed that the recent high growth rates of China and India have checked the trend in world wide, international – i.e. between-countries – income inequality; nonetheless, the situation of many African countries has become much worse. Moreover, the economic inequality within single countries seems to have increased for many countries and particularly the developed ones.

Atkinson and Brandolini (2004) in a study of global world inequality which takes account of various conceptions and measurements of inequality conclude the following:

> When we adjust for the within-country distribution of income, the evidence is almost unequivocally of a rise in income inequality from 1970 to 1992, whatever the underlying conception of inequality. (2004: 17)

Let us look specifically at within-country inequalities. Tony Atkinson, a world expert on the measurement of inequality, at a lecture at La Sapienza University gave comparisons for the Gini coefficient of inequality between 23 developed or intermediate countries. It shows that around the year 2000 the within-country inequality in income distribution was highest in: Mexico (where the coefficient was almost 50 percent), Russia, USA, Estonia, Israel, UK and Italy; it was lowest in: Slovak Rep., Finland, Netherlands, Slovenia, Belgium, Norway, Sweden, Check Rep. and Germany. As regards the UK, the presentation gives further details that show that the Gini coefficient has increased from around 26 percent

---

[13] The increase in insecurity was suggested by C. Howson during the discussion at the Bologna workshop.

in 1977 to over 35 percent in 1990 with a slight decline to around 33 percent in the mid-1990s and a slight increase after that to reach the previous level of around 36 percent in 2000.

Atkinson (2003) traces the secular trend in the incomes of the top 1% UK earners during the XXth century. His Fig. 2:36 shows that the share of the top 1 percent declined steadily from after WWII to 1979 and then increased steadily during the Conservative and New Labour governments. The UK trend in the last 25 years is similar to that in the US – though the two shares in 2000 are respectively 13 and 17 percent – but not to that of France where the share has remained fairly constant throughout the period. One of his conclusions on the UK is that "... the shares of top incomes are now broadly back where they were at the end of the Second World War. The last quarter of a century saw an almost complete reversal of the decline in observed inequality at the top that had taken place in the preceding twenty-five years" (pp. 22-3).

Callinicos (2001) also reports that inequality has widened under New Labour and writes:

> ... during Blair government's first two years in office ... The richest 10 per cent of the population saw their income rise by 7.1 per cent, compared to only 1.9 per cent for that of the poorest 10 per cent. (2001: 52)

He thinks that the responsibility for such trends under New Labour lies largely with the "... shifting the fiscal burden from direct to indirect taxation" (Callinicos 2001: 53) started under the Thatcher government and continued under Blair-Brown. This strategy makes the overall system of taxation more regressive.

Regarding the EU – 15 countries – Gallino (2005: 166-7) reports that the percentage of incomes from wages and salaries in the total GDP went from 76 percent in 1976 to 68.5 percent in 2002. Similar figures for Italy show a share of 76.6 percent in the period 1975-79 against 62.2 percent in 2000-03.

Italy is the country that de Finetti would have been most interested in. Here we have evidence from two papers from the Banca D'Italia. Brandolini et al. (2004: paper n. 530) deals with wealth distribution in Italy in the 1990s. Their Table 6 (p. 31) shows that the bottom 40 per cent of the population in terms of wealth per capita, has seen its share of wealth decline from 8.3 to 7.0 between 1989 and 2000; the next 40 per cent has also experienced a decline in the same period: from 33.8 to 29.2 per cent. To whom has the extra wealth gone? To the richest people. In fact, during the same period, we see the following changes: the top 10 per cent of the population have increased their share of national wealth from 40.2 to 48.5 per cent; the top 5 per cent of the population had an increase from 27.3 to 36.4 per cent; the top 1 per cent – the very, very wealthy Italians – have increased their share from 10.6 to 17.2 per cent. Between 1989 and 2000, the Gini coefficient for wealth distribution has increased from 0.553 to 0.613.

The comparisons between these figures and those for the same coefficient related to income, show that – as for other countries – wealth distribution tends to be much more concentrated than income distribution. Brandolini *et al.* (2004: 41) conclude on this point:

> The distribution of wealth is a lot more unequal than the distribution of income. In 2000 the Gini index was 0.61 for net wealth, compared with 0.37 for disposable income; it was 0.60 for tangible assets, and a much higher 0.81 for financial assets.

Brandolini *et al.* (2001: paper 427) in a study of trends in salaries and wages between 1977 and 1998 reach the following conclusions.

> Between 1977 and 1989, both mean and median real monthly net earnings rose by about one fourth or 1.8 per cent per year … In the following nine years, they declined by around 1 per cent per year; some of this reduction was due to the spread of part-time work … In the 1990s, the decline in monthly wages was compounded with a reduction in the average number of months worked … causing a pronounced fall in annual income from employment … much of the fall in net earnings in the 1990s may have been caused by the rising fiscal burden".

This trend in earnings led to an increase in the number of low paid and poor people. The same paper concludes on this point:

> … the diffusion of low-paid jobs evolved in parallel with that of earnings inequality. The proportion of low-paid workers declines from 17 per cent in 1977 to a minimum 8 per cent in 1989, rose to 16 per cent in 1993, and after a fall in 1995, reached a peak of 18 per cent in 1998. (Brandolini *et al.* 2001: 44)[14]

The rise in part-time work accounts for this pattern. The authors' further conclusion on poverty is that:

---

[14] Low-paid workers are defined by OECD (1996: 69) as "… workers who earn less than two thirds of median earnings of all full-time workers".

... the probability of being in poverty is more closely correlated with the amount of employment in the household, particularly employment of members other than the head, than with low pay. (Brandolini *et al.* 2001: 44)

There is no doubt that inequality has increased world wide since de Finetti was writing. At the time, he thought that a Martian would be appalled at the inequalities on Planet Earth (1976: 19 cited in Section 2); what would he think today?

3.2 Changes in economics

As I mentioned above de Finetti strongly criticizes the neo-classical economic theory on the basis of the various issues I have considered. Now, when de Finetti was writing it is fair to say that the accepted paradigm in economics was Keynesianism and that macro economics was the most relevant approach to the study of the economy. Following the stagflation period of the 1970s and other economic, social and political developments, Keynesianism – which is a variant of what Keynes said and advocated – was criticised and gradually a new orthodoxy was developed. The latter is based largely on microeconomics and on a stronger version of neo-classical economics: free market, acceptance of the status quo regarding the distribution of resources; supply side rather than demand side policies and thus emphasis on flexible labour markets. Partly thanks to the new theories, casualization of labour, short term contracts, high levels of unemployment have been and are widely accepted, justified and seen as the fault of workers and trade unions.

A more recent development is the large number of critiques of the new orthodoxy including the critique of excessive mathematization and abstraction of much economics (Lawson 1997; Gillies 2004). As in other historical periods and on other issues, a revolt on this particular point started in France in 2000 and has led to a lively debate with involvement by economists and non economists.[15] However, it has had very little impact on the practice of the teaching and on the economic profession so far.

Nonetheless, there are also critiques of unbridled liberalization and free market economics.[16] For example Roncaglia (2005: 121) concludes his book with the following statement which de Finetti might have agreed

---

[15] An account of the revolt and the following debate is in Fullbrook (2003).
[16] See various contributions in Fullbrook (2004).

with: "[the market] is not a *deus ex machina*, an invisible hand in which to trust passively, but a complex institution, within which there is a very large space for deliberate political intervention, an intervention directed towards the twin goal of justice and freedom."[17]

Moreover, there is a relatively new development in the social sciences that has some affinity with de Finetti's position and that might have appealed to him, at least in some aspects: "happiness" studies to which I now turn.

4   "Happiness" studies and de Finetti's views on economics

The last decade has seen a surge of studies on "happiness". The reason why I consider them here is that they have a bearing on de Finetti's writings on economics. In my view, these studies are indeed a development which he might have been interested in and we may even say that he was a precursor of some of the ideas behind "happiness" studies. My notes on "Happiness" come mainly from a reading of Richard Layard (2006, first published in 2005).

4.1   Layard on "Happiness"

The study of "happiness" may, at first, appear as a strange area for economists to delve in; is it not more the realm of psychology? The origin of this "New Science" – as Layard labels it – is, in fact, in psychology and in neuroscience. Recent diagnostic tests have shown that the brain does indeed record feelings of happiness or misery (Layard 2005: Ch. 2) and this is consistent with the results of survey into peoples' feelings and moods as well as circumstances leading to those feelings. So happiness has an objective dimension and some researchers talk of a "new science".

The study is multidisciplinary and indeed it is one of the strengths of Layard's book that it brings together many threads of thought – from neuroscience to psychology to sociology to politics to economics – and blends them together to make a coherent story with relevant policy implications. In Chapter 5 titled: "So what does make us happy?" Layard gives us seven factors affecting happiness: family relationships; financial situation; work; community and friends; health as well as personal freedom and personal values. However, though the absolute level of income matters for the happiness of the very poor and poor, as we grow richer other factors become more relevant: how well off we are comparative to others;

---

[17] My translation from Italian.

how well related and supported we are within the family, friends, community and work environment; how healthy we are. The major source of unhappiness is mental illness which affects many individuals and whose effects spill over to family, friends, and communities.

How does economics come into all this? The economist comes in first in terms of the possible questioning of economics' goals in the light of results on indicators of happiness. In the US there has been a steady increase in incomes per capita since the mid 1960s; however, there has not been increase in the percentage of people who declare themselves very happy. The story in Britain and other developed countries is similar. The overall results show that increases in incomes make people happier at earlier stages of development; when high levels of incomes per capita are reached people become happier – temporarily – only when their income is higher than that of other people around them. This leads to feelings of happiness if income distribution is in one's favour; however, this makes the relatively poorer people unhappy. So it is a zero-sum game or indeed a negative-sum game if the number of people who lose out exceeds the number of those who gain, as is usually the case.

Layard points out that there are "happiness" externalities from income distribution: if X is the only employee to receive a rise in salary or bonus, everybody else becomes less happy then they were before. Thus inequality does matter and Layard maintains that performance-related pay increases unhappiness by singling out individuals against groups. The rat race makes people unhappy; even those who win are happy only temporarily and they will soon want more. If everybody is better off, the rat race makes them all miserable: it is like behaviour in a football stadium: if one person stands while everybody is seated s/he can see better. However, this behaviour may prompt other to stand up and eventually nobody is better off.[18] Moreover, a culture of long hours of work generates unhappiness by preventing the development of good relationships within the family, community and friends. So the recommendation is that people should rebalance their work-life pattern, offer themselves for fewer hours and be contented with lower incomes. Moreover, as insecurity is also a great source of unhappiness, people must be prepared to trade off more security for lower wages.

Second, the economist comes in because the elements that seem to matter for people's happiness have a bearing on economics and on eco-

---

[18] As already mentioned de Finetti gives – on a similar vein – the example of people sitting or standing in theatres.

nomic policies. Economics is involved directly because of the impact on happiness of economic elements such as: wages, income distribution, working conditions. More indirectly the economic condition of individuals and countries affects the health of the population and its effects spill over into family members, friends and community.[19] Moreover, economics affects indirectly other "happiness" factors implicit in the following questions: how much time does our working life leave us to cultivate our personal relationships and nurture our families? How does income inequality shape our social relationships at work and outside it? To what extent are our personal and social relationships affected by the competitive environment in which we live?

A third route in which economics and the economist come in is in the development of policies to reach the aim of greater happiness or less misery. While the first part of Layard's book is devoted to a formulation of the problem and to bringing together evidence, the second part ("What can be done?") presents possible solutions.

Layard is in favour of progressive taxation not only to enable society to meet reasonable levels of public expenditure, but also as a means to a less unequal society and thus a less unhappy society.

4.2   De Finetti's likely views and "Happiness" studies

How does all this fit in with de Finetti's views on economics? There are indeed many analogies as well as differences between the two authors. Let us look at the analogies first.

−   Both de Finetti and Layard conceive an economics with wider aims than those set purely by the market: they want an economics with a human face, one that puts at the centre people rather than faceless markets. Layard wants the pursuit of happiness; de Finetti wants social justice and rational, efficient systems in the private and public spheres, systems that help rather than harass the individual.

−   Both authors express concerns for individuals, their emotions and well being; both favour a micro approach with individuals at the centre of analysis and policies.

---

[19] Marmot (2004)'s extensive research into the relationship between social status and health points to the fact that the rat race and excessive competition in the work place increase the probability of diseases for the less successful ones; the conclusion seems to be that, independently of the level of income per capita, inequalities and the rat race culture are killers.

- De Finetti's work as much as Layard's shows dissatisfaction with the standard macro indicators in economics such as GDP. For Layard this is due to the fact that GDP does not reflect peoples' happiness; for de Finetti it is because GDP does not take account of externalities and of the production of useless or harmful products; neither does it account for inequalities and social injustices.
- Externalities which play such a strong role in de Finetti, writings, are considered though to a lesser degree by Layard as well.
- Both de Finetti and Layard see the need to make society less unequal: de Finetti as a result of his passion for social justice; Layard in search of a less unhappy society.
- They both want a less insecure society.
- They are both critical of societies that are too competitive and lead to the rat race.
- Both consider the issue of trust to be important in the organization of society. The pursuit of happiness (Layard) and the aim of a more just and rational society (de Finetti) should put trust and cooperation centre stage.
- They are both in favour of a multidisciplinary approach. De Finetti uses his mathematical skills to show the feasibility of solutions to the problem of taking account of peoples' preferences in the sphere of ordinary choices as well on choices that involve political judgements. Layard uses results from psychology, neuroscience and sociology to show that objective measures of wellbeing other than the standard GDP are possible.
- Both authors take the capitalist system as given and permanent but would like to see changes in priorities as well as (de Finetti) in institutions.

On the issue of trust[20] it should be noted that the level of mistrust in society may have increased partly as a result of increased individualistic values but in part it may have been generated by the policies and behaviour of governments. The practice of continuous assessment and of set-

---

[20] It is interesting to see that while levels of trust seem to have declined, studies on trust and its relevance for businesses have mushroomed. Many of these studies – particularly in the international business literature – look at the use of the concept and the practice in Asian countries. This is partly connected with the fact that the higher levels of outsourcing in the economy need higher levels of trust between principal and contractor: it is as if the level of trust has plummeted at a time when developments in the world of production require more of it.

ting targets shows a high level of mistrust which – far from leading to better performance – leads to opportunistic behaviour: if the employer does not treat teachers, doctors, lecturers etc. as professionals, they may stop behaving as such; they will then direct their efforts to meeting targets[21] – whatever they may be – rather than to provide a service as professionals. Layard thinks that this situation leads to more unhappiness; de Finetti would have seen the inefficiency side in this system of controls because of: (a) the cost of carrying out the evaluations and monitoring; (2) the distortions caused to employees' behaviour.

Yet there are major differences between these two authors.

–   There is in de Finetti a deeper questioning of the market system and its alleged efficiency. Layard sees the market mechanism as a source of efficiency while de Finetti sees large inefficiencies in the production of useless or harmful products.
–   There is more passion in de Finetti about issues of social justice.
–   Moreover, de Finetti thinks that the shortcomings of capitalism are obvious and most people would agree on the need to correct them provided they are honest, not hampered by prejudice and vested interests and are put in a position to fully understand the issues.

A specific criticism that could be levelled at Layard's position is the following. His points about the excessive amount of competition among people, the destructive rat race, the long working hours with their negative effects on families are all too real. However, when it comes to remedies and policies he seems to put all the emphasis on the part of individuals with the implication that individuals have choices on how many hours they work and on whether to compete for positions or not. There is no attempt to see the system as a whole and to see that business and the market system have a (the?) major role in the structure and condition of the labour market and thus in whether jobs are secure or insecure or how many hours we should be working and thus what sort of life-work balance we end up with.

Layard sees the rat race as an innate trait in humans linked to the survival of the fittest. This approach sits at odds with his well documented positive effect that cooperation with – and trust in – other human beings has on happiness. It is possible that the main driver of the culture of status and the rat race is business and the market system; in this case it is

---

[21] A good discussion of the problems created by the culture of monitoring, assessing evaluating on British Universities is in Evans (2004).

the inner working of capitalism that must bear most of the responsibilities for the problems that lead to unhappiness; the problems deriving from imbalances in work-life patterns cannot be laid only at the door of individuals. Moreover, by focusing on individuals – with some responsibility on governments to ease the pains – we risk putting extra burden on them. Individuals are made to feel responsible for problems that are inherently social in nature. Ordinary people may end up more unhappy as they are made to feel something along the following: "not only I cannot win the rat race, neither can I be satisfied with my situation in life, or plan a more successful work-life balance and generally be at peace with myself: I am a total failure on all counts!"

Do these criticisms apply also to de Finetti's vision? He also accepts the capitalist system as the natural system for organizing economic life. However, he is much more critical of this system and prepared to be more interventionist to correct short comings which he sees as completely obvious and transparent particularly in terms of inequality and social injustices. He puts the burden of responsibility on the system and its institutions as well as on those individuals with power within it and not on ordinary individuals.

5   Summary and conclusions

This chapter tries to give a flavour of de Finetti's work in economics. After a brief exposition of my reading of his views, I tackle the issue of how his views on the economy, society and economics might be even more applicable today then in his own days. I do this by discussing the changes in the economy as well as in economics that we have witnessed in the last 25 years. This leads to a consideration of "Happiness" studies and their possible relationship (analogies and differences) with de Finetti's approach to economics.

Imagining a different course of history is always difficult and probably not very helpful in general. Nonetheless, while writing these notes I have often thought what might have been if de Finetti had been born 30 years later and would still be with us. Well, I think it would have been very nice to have him at the celebration of his 70th birthday. However, I am also convinced that he would be a very unhappy man to see how the society and economy on which he had so many aspirations and towards whose improvement he had devoted so much energy has changed for the worse, in the opposite direction to his "utopian" vision in almost all dimensions.

I would like to finish by touching on two further points. The first point relates to a possible solution of the *"enigma"* posed by de Finetti's approach to economics. De Felice (1981) as a conclusion to his introduction to de Finetti's works, writes that he agrees with Carnap that de Finetti is a *"puzzle"*. De Felice gives this word its board game meaning as jigsaw puzzle and sees de Finetti's works as a series of pieces that fit together neatly and coherently. This is an interesting interpretation and one which I fully endorse: indeed it applies very well to the relationship between de Finetti's views on economics and those on probability. It also applies to the relationship between his views on the application of mathematics to economics (the usefulness criterion) and his philosophical views on how we should choose concepts (their usefulness) (de Finetti 2006: 127).

However, I would like now to consider a further meaning of the word *puzzle*: puzzle as "enigma/mystery".[22] As already mentioned de Finetti's approach is one that starts from the individual person as consumer and citizen; this is in accordance with his subjective approach to probability. Yet in economics he is the most passionate advocate of social issues and many of his pronouncements relate to issues of equity and social justice. How is this "social" approach compatible with his individualistic approach? Again the solution may be found if we look at his approach on probability and compare it with his approach in economics. The probability conumdrum – consistency of the subjective approach with the axioms of probability – was solved by imposing consistency between bets to avoid a Dutch Book. In economics, his determination that economics should be *normative* and that people should express their preferences individually, transparently and directly led to his modified Pareto system and away from a macro economics approach. Using the same methodology as in his probability theory, he worked on a system in which the preferences of individuals in the realm of social choices were to be made consistent within the individual's overall set as well as compatible with the desiderata of other individuals in society. Thus he uses his philosophical approach to probability and his mathematical skills to work towards an economics with a human and social-justice face: one in which the citizen as social individual comes first; the scientists help to overcome inconsistencies; and the state intervenes to set priorities and to secure the achievement of people's desiderata in the individual and social spheres.

---

[22] I should note that, most likely, none of these interpretations of *puzzle* corresponds to what Carnap had in mind. As D.A. Gillies has pointed out to me, Carnap may have found puzzling that de Finetti should have come out with a non objective view of probability.

The second point refers to the language of de Finetti's writings. The fact that his works are full of passion made it more enjoyable for me to read them but more difficult to express in my own words the full impact they have on the reader. His language is not easy to follow: he often writes long sentences – as in many Italian writings – full of qualifications and adjectives designed to express all the possible feelings and angles on the issue under discussion.

Moreover, there are many "deFinettinisms" (my own expression): these are words specific to de Finetti and which are not to be found in any vocabulary. They are made up by him; they are his way of expressing his strong feelings. When feelings run high we are sometimes left speechless: our language is not strong or precise enough to express those feelings and thoughts. Those of us used to more than one language now and then experience that some feelings and expressions are better rendered in a different language from the one we are speaking at the time; often this is our native dialect which may point to the fact that the language we best express ourselves in is linked to our childhood and to the way we used to express our feelings when we first learned to speak.

This point is made explicit in some of the works of Luigi Meneghello, a writer and former Professor of Italian at the University of Reading in Britain. He is fluent in Italian and English but always preferred to resort to his native Vicentino – dialect from the province of Vicenza – to express his more subtle or strong feelings; indeed he says that he found it easier to translate/express English poetry in Vicentino rather than in Italian. In my view this is unlikely to have anything to do with the richness of Italian versus Vicentino; it may be related to the fact that the strong feelings of poems may lead the author/translator to use the language of his strong and subtle feelings: in the case of Meneghello such a language was indeed the Vicentino dialect. For de Finetti the feelings are sometimes so strong that he has to construct his own words.

Appendix

The three books by de Finetti used in the above discussions are collections of essays written at different times. It may be useful for the reader to be able to identify the year when certain statements and ideas by de Finetti where first put forward. To this end I here report details of the papers published in the three books by de Finetti referred to above. The following details are given: (1) Chapter, pages and title of each of the papers written by de Finetti himself; (2) full reference to the first date of publication; (3) indication – as given by de Finetti himself in the synopsis to the relevant chapter – of whether the later publication contains amendments. If no details regarding (2) and/or (3) are given it means that the paper was written expressly for the volume or that no amendments were reported by de Finetti.

De Finetti, B. (1969), *Un matematico e l'economia*, Milano: Giuffrè

Capitolo primo, "All'attacco contro i feticci", pp. 15-32.

Capitolo secondo, "Il tragico sofisma", pp. 33-49. First published in *Rivista Italiana di Scienze Economiche*, Anno VII, fasc. III, maggio-giugno 1935.

Capitolo terzo, "Formulazione matematica dei controesempi", pp. 50-66.

Capitolo quarto, "L'incertezza", pp. 67-76.

Capitolo quinto, "Nuove prospettive", pp. 79-91.

Capitolo sesto, "Benvenuto al disgelo", pp. 92-104. First published in *Civiltà delle Macchine*, Anno X, n. 3, maggio-giugno 1962 with the title (which de Finetti tells us was given by the editors of the journal in substitution to his own): "Teorie e realtà nell'economia europea".

Capitolo settimo, "La teoria dei giochi: cosa ci dice; su cosa ci invita a riflettere", pp. 105-143. First published in two instalments in *Civiltà delle Macchine*, Anno XI, nn. 4 e 5, luglio-agosto e settembre-ottobre, 1963 with the following title: "La teoria dei giochi" and "Riflessioni attuali sulla teoria dei giochi".

Capitolo nono,[23] "Econometristi allo spettroscopio", pp. 174-188. First published in *Rivista Trimestrale*, n. 15-16, sett.-dic. 1965.

Capitolo decimo, "Nuovi aspetti", pp. 191-196.

---

[23] Capitolo ottavo was written by Dario Furst in 1963.

Capitolo undicesimo, "L'apporto della matematica nell'evoluzione del pensiero economico", pp. 197-229. First presented at the VII Congresso dell'Unione Matematica Italiana, Genova 30 sett.-5 ott. 1963; published later as a separate paper in the series by Edizioni Ricerche, Roma and in 1964 in the *Atti* del Congresso, ed. Cremonese. Small amendment: a note has been added to the 1969 edition.

Capitolo dodicesimo, "Unità di concetti e varietà di metodi nella Ricerca Operativa", pp. 230-244. First published in *Mondo Economico*, 14-XI, 1959.

Capitolo tredicesimo, "Sulle applicazioni della Ricerca Operativa", pp. 245-251. First published in the volume *I sistemi aziendali avanzati*, ed. I.R.I., 1966. Some amendments.

Capitolo quattordicesimo, "Osservazioni sulla programmazione", pp. 252-260. First presented at the "Convegno di studi sui modelli di programmazione nei paesi della Comunità Economica Europea", Firenze, 7-9 giugno 1965; published in the Proceedings *Modelli Econometrici per la Programmazione* edited by G. Parenti, Scuola di Statistica dell'Università, Firenze, 1965, pp. 435-39.

Capitolo quindicesimo, "La teoria delle votazioni", pp. 261-269. First published in *Rivista di Diritto Finanziario e Scienza delle Finanze*, 1959.

Capitolo sedicesimo, "Sicurezza sociale e obiettivi sociali", pp. 270-293. First presented as "Sviluppo della popolazione e sicurezza sociale" at the "Convegno di studi sui problemi attuariali e statistici della Sicurezza Sociale", Roma, 18-21 giugno 1956, published in "Atti del Convegno" ed. Ordine Naz. Degli Attuari, Roma, 1956. Amended.

Capitolo diciassettesimo, "Disinfestazione antiimbecillistica", pp. 294-328. First published in *Homo Faber*, n. 160, 1965. Amendments: a section derives from a speech given in 1966; a short conclusion was written in 1968.

De Finetti, B. (a cura di) (1973). *Requisiti per un sistema economico accettabile in relazione alle esigenze della collettività*, Milano: F. Angeli.

Most of the material for this book was presented at Summer Schools between 1971 and 1973.

De Finetti, B. (1976). *Dall'utopia all'alternativa*, Milano: F. Angeli. This book contains contributions by several authors. The one by de Finetti is: "Dall'utopia all'alternativa" (1971-1976), pp. 7-51.

References

Acocella, N. (1998). *The Foundations of Economic Policy. Values and Techniques.* Cambridge: Cambridge University Press.

Acocella, N., Ciccarone, N., Franzini, M., Milone, L. M., Pizzuti, F. R. and Tiberi, M. (2004). *Rapporto su povertà e disuguaglianze negli anni della globalizzazione.* Napoli: Edizioni Colonnese l'ancora del Mediterraneo, Pironti.

Adriani, R. (2006). "Bruno de Finetti e la geometria del benessere". Dipartimento di Scienze Economiche, Università di Pisa, Mimeo.

Atkinson, A.B. (2003). "Top Incomes in the United Kingdom in the Twentieth Century". Mimeo, December.

Atkinson, A.B. (2006). "Income Inequality in Rich Countries". Slides, April.

Atkinson, A. and Brandolini, A. (2004). "Global World Inequality: Absolute, Relative or Intermediate?". Mimeo, preliminary version.

Brandolini, A., Cipollone, P. and Sestito, P. (2001). "Earnings Dispersion, Low Pay and Household Poverty in Italy, 1977-1998". Banca D'Italia, *Temi di discussione del Servizio Studi*, n. 427.

Brandolini, A., Cannari, L., D'Alessio, G. and Faiella, I. (2004). "Household Wealth Distribution in Italy in the 1990s". Banca D'Italia, *Temi di discussione del Servizio Studi*, n. 530.

Callinicos, A. (2001). *Against the Third Way.* Cambridge: Polity.

de Felice, M. (1981). "Prefazione" in B. de Finetti, *Scritti (1926-1930)*. Padova: Cedam, p. XXIX.

de Finetti, B. (1969). *Un matematico e l'economia.* Milan: Giuffrè.

de Finetti, B., ed. (1973). *Requisiti per un sistema economico accettabile in relazione alle esigenze della collettività.* Milan: F. Angeli.

de Finetti, B. (1976). *Dall'utopia all'alternativa.* Milan: F. Angeli.

de Finetti, B. (2006). *L'invenzione della verità.* Milan: Cortina. Submitted to the Reale Accademia d'Italia in 1934.

Evans, M. (2004). *Killing Thinking. The Death of the Universities.* London and New York: Continuum.

Frenz, M. and Ietto-Gillies, G. (2007). "Does Multinationality Affect the Propensity to Innovate? An Analysis of the Third UK Community Innovation Survey". *International Review of Applied Economics* 21, 1: 99-117, January.

Friedman, M. (1953). *Essays in Positive Economics.* Chicago: University of Chicago Press.

Fullbrook, E., ed. (2003). *The Crisis in Economics*. London: Routledge.

Fullbrook, E., ed. (2004). *A Guide to What's Wrong with Economics*. London: Routledge.

Gallino, L. (2005). *L'impresa irresponsabile*. Milan: Einaudi.

Gillies, D. A. (2004). *"Can Mathematics Be Used Successfully in Economics?"*. In *A Guide to What's Wrong with Economic*, ed. by E. Fullbrook. London: Routledge, Ch. 18, pp. 187-97.

Gillies, D. A. (2006). "Lessons from the History and Philosophy of Science Regarding the Research Assessment Exercise". In *Philosophy of Science*, ed. by Anthony O'Hear. Cambridge University Press; also at www.ucl.ac.uk/sts/gillies.

Ietto-Gillies, G. (2002). *Transnational Corporations. Fragmentation amidst Integration*. London: Routledge.

Ietto-Gillies, G. (2005). *Transnational Corporations and International Production. Concepts, Theories, Effects*. Cheltenham UK: Edward Elgar.

Lawson, T. (1997). *Economics and Reality*. London: Routledge.

Layard, R. (2005). *Happiness. Lessons from a New Science*. Penguin Books.

Marmot, M. (2004). *Status Syndrome. How Your Social Standing Directly Affects Your Health*. London: Bloomsbury.

Musgrave, A. (1981). "Unreal Assumptions in Economic Theory: the F Twist Untwisted". *Kyklos* 34 (3), pp. 377-87.

Meneghello, L. (2005). *La materia di Reading e altri reperti*. Milano: BUR.

Organization for Economic Co-operation and Development (OECD) (1996). "Earnings Inequality, Low-Paid Employment and Earnings Mobility". *Employment Outlook*, pp. 59-108. Paris: OECD.

Rapoport, A. (1962). "The Use and Misuse of Game Theory". *Scientific American*, Dec., pp. 108-118.

Roncaglia, A. (2005). *Il mito della mano invisibile*. Bari: Editori Laterza.

Samuelson, P. A. (1963). "Comment on Ernest Nagel's 'Assumptions in Economic Theory'". *American Economic Review*, May. Reprinted in *The Collected Scientific Papers of Paul A. Samuelson*, ed. by J.E. Stiglitz (1966). Cambridge, MA: MIT Press, vol. 2, no. 129, pp. 1772-8.

Tiberi, M. (2006). "Gli insegnamenti economici". In *La Facoltà di Economia. Cento anni di storia*. Rome: Rubettino editore, pp. 335-428.

Zanfei, A. (2000). "Transnational Firms and the Changing Organization of Innovative Activities". *Cambridge Journal of Economics* 24, pp. 515-42.

10

On Some of de Finetti's Unknown and Unsung Contributions to Statistics

Eugenio Regazzini [*]

This paper aims at commemorating the centenary of the birth of Bruno de Finetti (Innsbruck, 13 June 1906). It links a few remarks on the power of exchangeability in detecting inconsistencies in the conventional Bayesian paradigm with the analysis of a couple of papers belonging to the least known part of de Finetti's impressive scientific corpus. The former shows de Finetti's decisive contribution to the so-called fundamental theorem of mathematical statistics so far known as *Glivenko-Cantelli theorem*. The latter contains a concise treatment of the elementary problem of linear correlation, all the more informative, though as it has a wealth of original cues conducive to further research.

From the very title of this paper it is clear that I do not intend it to be a mere review of Bruno de Finetti's life and work. Neither does it claim to be exhaustive in any respect. Somewhat pointedly, I meant it to draw attention to some of his specific contributions to Statistics that might have passed unnoticed although they proved to be excitingly seminal. I hope this might be a proper way both to (implicitly) disclose some hidden aspects of his character, and to acknowledge priority of his work with respect to some key probabilistic and statistical topics.

Before getting to the heart of the specific topics I intend to deal with, let me just mention a few important data of de Finetti's biography.

[*] This paper was financially supported by MIUR grant 2006/134 526. It reproduces, with minor changes, a lecture given at the XLIII Meeting of the *Società Italiana di Statistica*.

1   Some essential biographical notes

In 1923 he enrolled in the engineering faculty of the *Politecnico di Milano*. During his third year of study he initiated his first independent mathematical research. In particular, Bruno had been attracted by probability and statistics indirectly through the reading of a paper on Mendelian heredity by Carlo Foà, a biologist. By means of a system of ordinary differential equations, he was able to describe the evolution of the frequencies of the homozygotes and of the heterozygotes and to state the conditions under which they converge to a limit. His results were appreciated by Foà, Giulio Vivanti, a mathematician well-known for some noteworthy contributions to complex analysis, who at that time taught mathematical analysis at the university of Milano, and by Giorgio Mortara, a statistician active in the same university. Mortara submitted part of de Finetti's research for publication in *Metron* to Corrado Gini, editor of that journal, who published the paper in an issue of 1926. Gini was so favorably impressed by this engineering student that he ensured him a position, straight after he managed to get a degree, at the *Istituto Centrale di Statistica* he presided over at that time. Bruno made the most of this promise to persuade his reluctant mother to comply with his desire to shift engineering to mathematics. That save him one year of study (engineering was a five-year course, whilst mathematics a four-year course) but, above all, gave him the opportunity to follow his more theoretical bent. Thus, he obtained a degree in applied mathematics in 1927 by discussing his graduation dissertation with Giulio Vivanti. Soon afterwards, de Finetti accepted the job at the Istituto Centrale di Statistica in Roma. He worked there until 1931. In those years, he laid the foundations for his principal contributions to probability: definition and analysis of sequences of exchangeable events; definition and analysis of processes with stationary independent increments. Moreover, in 1930 he qualified in a competition as a university lecturer of Mathematical Analysis (members of the board of examiners were: Giuseppe Peano, Mauro Picone and Salvatore Pincherle).

In 1931, he moved to Trieste to hold a position as actuary at the *Assicurazioni Generali*. Bruno's father and grand-father had worked there as engineers. During his stay in Trieste, he developed the research he had started in Rome, but he also obtained significant results in actuarial and financial mathematics as well as in economics. In spite of his impending actuarial activity, in Trieste de Finetti also started his teaching career (mathematical analysis, actuarial and financial mathematics, probability) with a few years's stint at the universities of Padova and Trieste. He won a nationwide competition in 1939 for a chair in Financial Mathematics, but it was only in 1946 that he obtained his chair as full professor at the faculty of Sciences of the university of Trieste. In 1951 de Finetti changed to the faculty of Economics at the same university. In 1954 he moved to the faculty of Economics of the university of Rome, "La Sapienza", where he stayed until 1961, when he changed to the faculty of Sciences, where he was professor of the Theory of Probability until 1976, the year of his retirement.

He died in Rome on 20 July 1985.

2 Exchangeability and Bayesian reasoning

In 1973, de Finetti was invited to give a lecture at the 39th Session (Vienna) of the International Statistical Institute. Under the title "Bayesianism: its unifying role for both the foundations and the applications of statistics", he dealt with a number of issues: statistical inference, decision theory, their frequency and non-frequency interpretations, inductive reasoning and inductive behavior. Part of his lecture focussed on the problem of how to get rid of the framework of "hypotheses" about parameters. In point of fact, along with statistical schemes (or models) where parameter θ is a factual, although unknown or hidden, quantity, there coexists a statistical modeling in which

θ is a merely fictitious, or mythical, pseudoentity, allowing to perform some reasoning "as if" it existed, but leading to absurdities would it be thought as really existing.

Some old-time demographers, in the age when probabilities where ordinarily conceived only with reference to urn-schemes, used to explain the mortality table with the image of the Parcae drawing each year a ball for each of us, to decide about life or death according to its color – white or black –, and using an urn where the fraction of black balls was increasing with the age.

But to imagine an urn with unknown but constant composition explaining at any drawing from the Pólya urn its outcome as resulting from the "hidden urn" would be even more difficult: much more artificial and preposterous than the plainly mythological picture of the Parcal. The "hidden urn" should in effect have, as its unknown but predetermined composition, the one which corresponds to the limit to which the composition of the Pólya urn should approach

through endless addition of new balls to the few put into it till now. (And it is almost sure that not even the Vestals would assure the continuation of such experiment for eternity, what would imply, incidentally, to get sometime more balls than atoms in the world; and, on the other side, there is no reason to expect such limit to exist, since stochastic, even if strong, convergence does not guarantee any conclusion on this point).
In such a situation, it is obviously only the predictive aspect (concerning the future outcomes – not the parameter!) that matters.

De Finetti pointed out that getting rid of the framework of "hypotheses" is actually the aim as explicitly expressed in Roberts 1965, who advocated that attention should be focussed on the predictive distribution concerning directly the quantities of interest rather than on the ones concerned with the parameters, which are nothing but auxiliary ingredients. He then mentioned that

> a particular case (but a particularly simple and important one) of Robert's distinction is the one where (as in the example just discussed) we have – in the terminology of Objectivists – independent events E_i (or random quantities X_i) with constant but unknown probability θ (or probability distribution, say F_θ). ... Under such assumption the events are *exchangeable* ... But also the converse holds ... That is what some Colleague ... called the de Finetti representation theorem. My aim was precisely the same as Roberts', if only restricted to the case of the most usual and important example of inductive reasoning (and, then, of inductive behaviour). The aim was to present induction as a very natural way of reasoning on probabilities of observable facts, avoiding metaphysical pseudoentities and obscurities.

These fragments from the ISI lecture do confirm in the most perfect possible way the thought expressed by their author more than forty years beforehand, and they give a clear-cut picture of the specificity of de Finetti's position in the Bayesian circle.

In fact, the problem he intended to deal with towards the end of the 1920th was, in a sense, of a more general extent, i.e.: "Can the subjective theory of probability provide a conceptually rigorous justification of probability assessments based on observed frequencies of analogous events?". A positive answer would imply that the subjective view, far from being incompatible with empirical approaches to probability, does comprise them as processes to evaluate probabilities when particular circumstances occur. In a complete probabilistic setting, these circumstances must be expressed through probabilistic models which translate the empirical idea of "analogous" events (or more general random elements) into a proper mathematical language. As a matter of fact, the idea of "repeatable events" which recurs both in the so-called frequentistic definitions of probability and in the explanation of objective approaches to statistical methods, boils down to the assumption that sequences of trials are carried out under analogous conditions. In a talk (see de Finetti 1932) given at the *International Mathematical Congress*, held in Bologna in 1928, de Finetti had introduced *exchangeability* as a probabilistic characterization of a *random phenomenon*, that is, a phenomenon which can be repeatedly observed under analogous environment conditions. He argued that a correct probabilistic translation of such an empirical circumstance leads us to think of the probability of k successes and $(n-k)$ failures in n trials as invariant with respect to which successes and failures alternate, whatever k may be (and, in the case of an infinite sequence of trials, whatever n may be). In other words, events $E_1, E_2, \ldots, E_N$ are said to be *exchangeable* if their law is the same as the law of $E_{\pi(1)}, \ldots, E_{\pi(N)}$ for any permutation π of $(1, \ldots, N)$; analogously, the elements of a sequence of events are called exchangeable when the previous condition obtains for $(E_i)_{1 \leq i \leq n}$ and for every n. For any sequence of events the problem whether the frequency of success can be used to estimate the probability of success has a precise formulation: for every pair of positive integers n, k, define the conditional probability $P(E_{n+k}|E_1, \ldots, E_n)$ and the frequency of success[1]

$$f_n(E_1, \ldots, E_n) := \frac{1}{n} \sum_{k=1}^{n} 1_{\mathbf{E_k}};$$

then, (a) check the deviation $|P(E_{n+k}|E_1, \ldots, E_n) - f_n(E_1, \ldots, E_n)|$, and (in the case of sequences) (b) verify whether it converges (stochastically) to 0 as the number n of trials goes to infinity.

One of the main achievements of de Finetti's research is the following theorem presented in 1928 and proved in de Finetti 1930:

If $(E_n)_{n \neq 1}$ is a sequence of exchangeable events, then

---

[1] For any event A, 1_A stands for its indicator, that is, $1_A = 1$ or 0 depending on A comes true or not.

(1) $\inf_{k,p\in\mathbf{N}} P\Big\{ \max_{n\le m\le n+p} |P(E_{m+k} \mid f_m(E_1,\dots,E_m))-$
$$f_m(E_1,\dots,E_m)| \le \epsilon \Big\} \ge 1-\eta$$

holds true for every pair of positive numbers ε, η, provided that n is larger than a suitable integer depending only on (ε, η).

Roughly speaking, one can say that in the case of sequences of exchangeable events, that is, in the case of analogous trials, one can legitimately use frequency as an estimator of the conditional probability of any future event E_{n+k}, given any outcome of $(E_1,\dots,E_n)$ yielding that particular frequency, provided that n is sufficiently large.

In order to avoid unessential problems of a purely technical nature, we intend to stick to the field of events and thus to try to explain de Finetti remarks already hinted at above about "hypotheses". As a matter of fact, it is easy to devise statistical problems in which attention must be focussed on prevision of frequency of N events, under the hypothesis one can get information frequency of n events with $n<N$. To this end, one can note that, for any $(x_1,\dots,x_n)$ in $\{0,1\}^n$ with $s_n=x_1+\dots+x_n$, in the case of exchangeable events,

$$P\Big\{ \mathbf{1}_{E_1} = x_1,\dots, \mathbf{1}_{E_n} = x_n, \sum_{i=1}^N \mathbf{1}_{E_i} = M \Big\}$$

$$= \frac{\binom{N-n}{M-s_n}}{\binom{N}{M}} P\Big\{ \sum_{i=1}^N \mathbf{1}_{E_i} = M \Big\} \qquad (M = s_n,\dots, N-n+s_n)$$

obtains and, by the Bayes theorem,

(2)
$$P\Big(\sum_{i=1}^N \mathbf{1}_{E_i} = M \mid \sum_{i=1}^n \mathbf{1}_{E_i} = s_n \Big)$$

$$= \frac{\binom{n}{s_n} P\{\mathbf{1}_{E_1} = x_1,\dots, \mathbf{1}_{E_n} = x_n, \sum_{i=1}^N \mathbf{1}_{E_i} = M\}}{P\{\sum_{i=1}^n \mathbf{1}_{E_i} = s_n\}}$$

where

(3) $P\Big\{ \sum_{i=1}^n \mathbf{1}_{E_i} = s_n \Big\} = \sum_{M=s_n}^{N-n+s_n} \frac{\binom{n}{s_n}\binom{N-n}{M-s_n}}{\binom{N}{M}} P\Big\{ \sum_{i=1}^N \mathbf{1}_{E_i=M} \Big\}$

De resent the initial segment of an infinite sequence of (exchangeable) events and N (e.g., size of statistical population) is immensely large compared with n (size of sample). The study of this realistic situation is based on the noteworthy result that the sequence $(\sum^n{}_1 1_{Ei}/n)_{n\geq 1}$ *mutually* converges in probability which, in turn, implies that the corresponding sequence of probability distributions $F_n(x):=P\{1/n \sum^n{}_{i=1} 1_{E_i\leq x}\}$, $x \in R$ and $n=1,2,\ldots$, weakly converges to a probability distribution function F, supported by $[0,1]$. Then, passing to the limit, as $N\to+\infty$, in (3),

$$
(4) \qquad P\left\{\sum_{i=1}^{n} 1_{E_i} = s_n\right\} = \binom{n}{s_n} \int_{[0,1]} \theta^{s_n}(1-\theta)^{n-s_n}\,dF(\theta)
$$

obtains for every n and $s_n=0,1,\ldots,n$. This is the celebrated *de Finetti's representation theorem*, described according to the original way of arguing of its discoverer. Consequently, the main problem of providing previsions for $\sum^N{}_{i=1} 1_{Ei}$ can be solved, besides (2), by

$$
(5) \qquad P\left(\sum_{i=1}^{N} 1_{F_i} = M \mid \sum_{i=1}^{n} 1_{F_i} = s_n\right)
$$

$$
= \binom{N-n}{M-s_n} \frac{\int_{[0,1]} \theta^M(1-\theta)^{N-M}\,dF(\theta)}{\int_{[0,1]} \theta^{s_n}(1-\theta)^{n-s_n}\,dF(\theta)}.
$$

Whence, (2) and (3) can be re-written, and simplified to a certain extent, as (5) and (4) respectively, and this shows that it is enough to assign the limiting distribution of the frequency of success, i.e. F, in order to master all the pertinent distributional aspects of an *infinite* sequence of exchangeable events. Then, although F can be considered as a very useful tool to specify those distributional aspects, it is but an "auxiliary" ingredient for the most statistical models. In spite of these incontrovertible remarks about the nature of what is generally called *initial* or a *priori distribution*, Bayesians persevere in formulating inferences on some random element "initially" distributed according to F, even when discovering any real empirical meaning for such a random element is clearly impossible. Under these unfortunately all too common circumstances, unknown parameters are but fictions, or mythical, pseudoentities, as recalled at the beginning of this section, and de Finetti's invitation mentioned therein, to get down to predictive aspects, does warn us against inconsistent arguments.

3 Empirical approximation of a probability distribution

Objectivists are used to study statistical procedures under an "extreme" form of exchangeability, that is, when observations $\xi_1, \xi_2, \ldots$ are supposed to be independent and identically distributed with an unknown distribution. If observations are real valued, it is customary to make reference to their common probability distribution function $x \mapsto F(x)$. In view of the assumption of stochastic independence, the conditional distribution for each future observation, given any number of past observations, must coincide with F. Then, the discussion and the study of relations between prevision of frequencies relative to future observations and observed frequencies (that, in the case of general exchangeable sequences, resort to statements of theorems like (1)) is based on the study of the behaviour of some distance between F and the *empirical distribution* of $\xi_1, \ldots, \xi_n$, that is

$$F_n(x) := \frac{1}{n} \sum_{i=n}^{n} \mathbf{1}_{(-\infty, x]}(\xi_i) \qquad (x \in \mathbf{R}, n = 1, 2, \ldots).$$

For a fixed x, the problem boils down to a simple application of the classical law of large numbers, but its complete solution entails more sophisticated tools from the theory of empirical stochastic processes (random functions whose realizations are frequency distribution functions).

In the first issue dated 1933 of *Giornale dell'Istituto Italiano degli Attuari (GIIA)*, a fundamental paper of Kolmogorov (see Kolmogorov 1933) appeared in which the great Russian mathematician gave the limiting distribution, as $n \to +\infty$, of $\sqrt{(n)} \|D_n\|_u$ where

$$\|D_n\|_u := \sup_{x \in \mathbf{R}} |F(x) - F_n(x)|,$$

under the hypothesis that F is *continuous*. This result, besides its groundbreaking value from a pure probabilistic viewpoint, influenced the development of statistical tests of goodness of fit and of the so-called two-sample problem.

In that very same *GIIA* issue of 1933, the paper immediately following Kolmogorov's was a paper by of Valeri Glivenko (cf. Glivenko 1933) in which it is proved, under the hypothesis of continuity for F, that $\|D_n\|_u$ converges to zero, with probability 1, when $n \to +\infty$.

This specific problem did attract de Finetti's attention. As a result he wrote a short paper, published in the third issue of *GIIA* (dated July 1933). See de Finetti 1933. As stated in the first section, the content consists of a few simple remarks concerning:

(a)   The way of proving the Glivenko theorem.

(b)   How to take the problem when the discrepancy between probability distribution functions is measured through a more suitable distance than $\|D_n\|_u$, i.e. Lévy's distance for arbitrary probability distribution functions F and G on $\mathfrak{R}$,

$$\lambda(F, G) := \inf\{\epsilon > 0 : F(x - \epsilon) - \epsilon \leq G(x) \leq F(x + \epsilon) + \epsilon \text{ for all } x\}$$

which metrizes weak convergence of probability distributions on R.

(c)   The study of the Glivenko problem, with respect to $\|D_n\|_u$, when the size of each jump of F represents the probability concentrated at the corresponding discontinuity point.

Before describing and discussing de Finetti's contribution with reference to Cantelli's better known paper (cf. Cantelli 1933), let me say a few words about point (c). Indeed, within the usual Kolmogorov axiomatic theory of probability, condition mentioned *sub* (c) is a necessary consequence of the axiom of σ-additivity. On the other hand, when simply additive probabilities are admitted, as is the case with de Finetti's theory based on the coherence principle, then the probability concentrated at any discontinuity point might be smaller than the jump, even vanish. This explains the rationale of (c).

Apropos of (a), de Finetti shows that the Glivenko theorem can be deduced, in an easy and elementary way, from Cantelli's classical strong law of large numbers (cf. Cantelli 1917)

which can be considered as source of inspiration for all these studies. [Translation of the underwriter]

As for (b), in Section 3 of de Finetti's paper it is shown that the restriction concerning the continuity of F is needless when the sup-norm is replaced by the Lévy metric, i.e.:

Whatever F may be, given $\varepsilon, \eta > 0$ there is $N = N(\varepsilon, \eta)$ such that

(6)
$$\sup_{p} P\left\{ \max_{N \leq n \leq N+p} \lambda(F_n, F) \geq \epsilon \right\} \leq \eta$$

which, within the usual σ-additive setting, is tautamount to asserting

$$\lim_{N \to +\infty} P\left\{ \sup_{n \geq N} \lambda(F_n, F) > \epsilon \right\} = 0 \qquad (\epsilon > 0).$$

When the circumstance mentioned sub (c) occurs, it is easy to show that the argument used to prove (6) for arbitrary probability distribution functions holds true in a slightly more restrictive sense, but basically just enough to extend Glivenko's result to discontinuous F.

Whence, de Finetti's remarks about the Glivenko problem highlight the very nature of problem itself and include the part of the theorem that is customarily attributed to Cantelli. Having ascertained that the name of de Finetti ought to appear, with good reason, among the discoverers of the so-called *fundamental theorem of mathematical statistics*, it is worthwhile examining things from Cantelli's vantage point. It should be recalled that Cantelli himself, as editor of *GIIA* (from 1930 to 1957), decided to accept de Finetti's paper. Moreover, it should be mentioned that his extension of the Glivenko theorem appears in the same July issue of *GIIA*, straight after de Finetti's paper. He quotes Glivenko's and de Finetti's papers, and it would pretty tough indeed to pinpoint what actually led to attribute the honour of the discovery to Cantelli and Cantelli alone. The only explanation could stem from Cantelli's hint that the content of the paper had actually be the subject of a few lectures given at the *Institut H. Poincaré* (Paris) in May of 1933. In point of fact, Cantelli's paper did appear, but two years later, in the *Annales* of that Institute, and covered the convergence of sequences of random elements. Cf. Cantelli 1935. It is split up into three parts, the first of which contains a detailed proof of the proposition formulated in his 1933 paper.

4 Correlation and concordance

The last paper I intend to describe deals with linear correlation between real-valued random variables. It was published in 1937 in the statistical supplement to a magazine of politics, history and economics expression of the so-called left wing of the fascist party. The title of the paper was *A proposito di "correlazione"* (de Finetti 1937) and can be traced back to a debate on the use and misuse of the Bravais-Galton correlation coefficient which came to the fore during the 23*rd* Session of the *International Statistical Institute* (London, 1934). Although the main goal of the paper is to seek an effective explanation of the essential properties of the coefficient and of its domain of application, its value is enhanced by a few original hints.

4.1 Geometry of correlation

As to the first point under scrutiny, de Finetti – well ahead of his times – suggests considering the following geometric interpretation. One considers the set D of all *random* numbers X, such that $E(X^2)<+\infty$, as a vector space, any vector being thought of as a class of random numbers which differ by a constant (with probability one). Then, denoting mean square deviation by $\sigma(.)$ and linear correlation coefficient between X and Y by $r(X,Y)$, the covariance $Cov(X,Y)=\sigma(X)\sigma(Y)r(X,Y)$ can be interpreted as *inner product* and, consequently, $\sigma(.)$ defines a norm on D. It turns out that $r(X,Y)$ is the cosine of the angle formed by the vector X with the vector Y, which is determined uniquely by setting $r(X,Y)=\cos[\alpha(X,Y)]$ *with* $0\leq\alpha(X,Y)\leq\pi$.

By starting from this geometrical interpretation, one easily grasps: the meaning of the extreme situations ($r = -1$, or $+1$), the meaning of positive (negative, respectively) correlation, and that of absence of correlation (i.e. orthogonality). Moreover, through such a representation, well-known properties of *inner product spaces* may remind us of geometrical images suitable for studying and solving various problems of a probabilistic or statistical nature. The following propositions, for example, can be seen as direct consequences of geometrical properties.

(A) If $X_1,\dots,X_n$ are linearly independent vectors (in the aforesaid sense), then they can be expressed as linear combinations of n random numbers $Y_1,\dots,Y_n$ satisfying: $E(Y_i)=0$ and $\sigma(Y_i)=1$ for every $i,r\,(Y_i,Y_j)=0$, for every $i\neq j$.

(B) If $X_1,X_2,\dots,X_n$, as in (A), satisfy $r(X_i,X_j)=\rho$ for every $i\neq j$ (pairwise equicorrelation) with $E(X_i)=0$ and $\sigma(X_i)=\sigma$ for every i, then ρ attains the permitted minimum value if and only if $\sum_{i=1}^{n}X_i=0$ with probability 1, in which case: $\rho = -1/(n-1)$ (Just think, in particular, to application to exchangeable random numbers.)

4.2 Classes of distributions with given marginals

Moving on to further original issues contained in the paper, we come across an amazing *ante litteram* argument involving distributions with fixed marginals, with which Hoeffding and Fréchet are generally credited. Cf. Hoeffding 1940 and Fréchet 1951, respectively. Consider events $E_1,E_2,E=E_1\cap E_2$ and their probabilities p_1,p_2 and p, respectively; set $q_i=1-p_i$ ($i=1,2$), $q=1-p$. Now, if p_1 and p_2 are *fixed*, then the range of the correlation coefficient r between (the indicator of) E_1 and (the indicator of) E_2 is, in general, a proper subset of $[-1,1]$. For instance, if $p_1<p_2$ and $p_1+p_2 \leq 1$, one has

$$-\left(\frac{p_1p_2}{q_1q_2}\right)^{1/2} \leq r \leq \left(\frac{p_1q_2}{q_1p_2}\right)^{1/2}.$$

Note that the two bounds coincide with −1 and +1, respectively, if and only if $p_1=p_2=1/2$. But, on the other hand, both these bound could be very close to 0 (when p_1 is close to 0). De Finetti extends these conclusions to pairs of random numbers X_1 and X_2 with assigned marginal distribution functions Φ_1 and Φ_2, respectively. Firstly, he argues that $r(X_1,X_2)$, in this case, can attain the value 1 (−1, respectively) if and only if $\Phi_1(x)=\Phi_2(ax+b)$ for every x and some $a>0$ and b ($\Phi_1(x)=1-\Phi(ax+b)$ for every x and some $a<0$, and b, respectively). Moreover, the range of r coincides with $[-1,1]$ if and only if $\Phi_1(x)=\Phi_2(ax+b)$ for every x, for some $a>0$, and Φ_1 is "symmetric".

At this point, de Finetti shows that, generally speaking, the extremes, say r_1 and r_2 of the range of $r(X_1,X_2)$ can be obtained

by considering the extreme cases in which the values of (X_1,X_2) lie on a decreasing (an increasing, respectively) curve, with probability 1. [Translation of the underwriter]

We see that this is the very same principle as the one that Fréchet will consider to define the extreme distribution functions of the classes that we routinely call *Fréchet classes*. In point of fact, de Finetti indicates how to determine the equations of the curves which support he extreme distributions and, consequently, obtains the following expression for r_1 and r_2 (when $E(X_i)=0$ and $\sigma(X_i)=1$ *for* $i=1,2$)

$$r_1 = \int_0^1 \Phi_1^{-1}(t)\Phi_2^{-1}(1-t)dt, \quad r_2 = \int_0^1 \Phi_1^{-1}(t)\Phi_2^{-1}(t)dt.$$

Finally, de Finetti uses this approach to clarify the nature of one of the problems which had sparked off the controversy mentioned at the beginning of this section, and to solve it. Reference is made to the relationship between stochastic independence and orthogonality (i.e., $r(X_1,X_2)=0$). He proves that: If Φ_1 and Φ_2 are assigned, then the two concepts turn out to be equivalent if and only if both the support of Φ_1 and of Φ_2 contains two points at the most.

4.3 Measures of concordance

With reference to the correlation coefficient, there was another question open to dispute, that is: Can this index be used to measure the *concordance* between X_1 and X_2?

Roughly speaking, two ordered characters are said to be concordant (discordant, respectively) if, being either character large, this indicates that the other character is large (small, respectively). According to this meaning of the term, due to Gini and studied by his School, concordance is more general than correlation, the latter corresponding to a very particular form of concordance or discordance (linear). Then, the measurement of concordance requires appropriate indices like the *indices of homophily* introduced by Gini (1915). In the last section of *A proposito di correlazione* de Finetti defines a new index of concordance/discordance, the one he considers as

the most simple and the most intrinsically revealing.

In order to point out the difference between the concepts at issue, he gives an example to show that the Bravais coefficient and this new index of concordance and discordance may have discordant signs. As to the definition of the new index for a (bivariate) distribution F, he considers two stochastically independent random vectors (X_1, Y_1) and (X_2, Y_2) with the same probability distribution function F. Clearly, the sign of the product $(X_2-X_1)(Y_2-Y_1)$ indicates concordance/discordance between two characters X and Y jointly distributed according to F, and an appropriate index can be obtained by taking the expectation of the signum of that product. This yields

$$C(F) = \int_{\mathbf{R}^4} \mathrm{sign}[(x_2 - x_1)(y_2 - y_1)] \, dF(x_1, y_1) \, dF(x_2, y_2).$$

It should be noted that if, instead of expectation of the sign of the product, we take the expectation of the product, then we obtain a concordance/discordance index which is undoubtably more related, than C, to r. Indeed,

$$\int_{\mathbf{R}^4} (x_2 - x_1)(y_2 - y_1) \, dF(x_1, y_1) \, dF(x_2, y_2) = 2\sigma(X_1)\sigma(Y_1)r(X_1, Y_1).$$

It's a plain fact that the sign of this index, unlike C, must be the same as the one of r.

Attentive readers have certainly noticed that C is but the index which, in statistical literature, is well-known as *Kendall's τ* (cf. Kendall 1938). Yet, de Finetti introduced this index one year before Kendall did, so that one might as well redesignate it as *de Finetti-Kendall index*. This had already been proposed in Cifarelli and Regazzini, but in vain (cf. Cifarelli and Regazzini 1996).

References

Cantelli, F.P. (1917). "Sulla probabilità come limite delle frequenze". *Rendiconti R. Accad. Naz. Lincei* V 26, pp. 39-45.

Cantelli, F.P. (1933). "Sulla determinazione empirica delle leggi di probabilità". *Giorn. Istit. Ital. Attuari* 4, pp. 421-424.

Cantelli, F.P. (1935). "Considérations sur la convergence dans le Calcul de probabilités". *Annales de l'Institut H. Poincaré* 5, pp. 3-50.

Cifarelli, D.M. and Regazzini, E. (1996). "De Finetti's Contribution to Probability and Statistics". *Statistical Science* 11, pp. 253-282.

de Finetti, B. (1930). "Funzione caratteristica di un fenomeno aleatorio". *Memorie della R. Accad. Naz. Lincei* VI, pp. 86-133.

de Finetti, B. (1932). "Funzione caratteristica di un fenomeno aleatorio". *Atti del Convegno Internazionale dei Matematici* (Bologna, 1928), vol VI. Bologna: Zanichelli, pp. 179-190.

de Finetti, B. (1933). "Sull'approssimazione empirica di una legge di probabilità". *Giorn. Istit. Attuari* 4, pp. 415-420.

de Finetti, B. (1937). "A proposito di 'correlazione'". *Supplemento statistico ai Nuovi Problemi di Politica, Storia ed Economia* 3, pp. 41-57.

Fréchet, M. (1951). "Sur les tableaux de corrélation dont les marges sont données". *Ann. de l'Un. de Lyon* 4, pp. 13-84.

Gini, C. (1915). "Di una misura della dissomiglianza tra due gruppi di quantità e delle sue applicazioni allo studio delle relazioni statistiche". *Atti R. Ist. Veneto di Sc. Lettere e Arti* 74, pp. 185-213. (Altre quattro Note apparse negli stessi Atti nel biennio 1915-1916).

Glivenko, V. (1933). "Sulla determinazione empirica delle leggi di probabilità". *Giorn. Istit. Ital. Attuari* 4, pp. 92-99.

Hoeffding, W. (1940). "Masstabinvariate Korrelations-theorie". *Schriffen Math. Inst. Univ. Berlin* 5, pp. 181-233.

Kendall, M.G. (1938). "A New Measure of Rank Correlation". *Biometrika* **30**, pp. 81-93.

Kolmogorov, A.N. (1933). "Sulla determinazione empirica di una legge di distribuzione". *Giorn. Istit. Ital. Attuari* 4, pp. 92-99.

Roberts, H. V. (1965). "Probabilistic Prediction". *JASA* 60, pp. 50-62.

11

Probability and the Logic of de Finetti's Trievents

ALBERTO MURA

1 Introduction

Philosophical research on conditionals is a complex topic. Since Ramsey, philosophers have noticed that the degree of assertability of an utterance of the form "if p then q" may be measured by its probability $\mathbf{P}(q \mid p)$. However, $\mathbf{P}(q \mid p)$ involves two propositions while natural language suggests that "if p then q" is a single proposition. Can we regard '|' as a genuine connective? Clearly, standard material implication is not suitable to represent '|', since from the laws of probability it follows that except in very special cases $\mathbf{P}(q \mid p) \neq \mathbf{P}(p \to q)$. Since there is no other decent Boolean candidate for conditional except material implication, it follows that '|' cannot be seen as a Boolean connective. David Lewis (1976) argued that it cannot be an iterable connective at all (let alone a Boolean one). Lewis's reasoning convinced many philosophers, like Adams (1975), that conditionals are not propositions (capable of being either true or false) and that asserting a conditional amounts to maintaining that $\mathbf{P}(q \mid p)$ has a very high value. In more recent times, researchers in the field of AI tried to challenge Lewis's result, by merging the Boolean algebra of ordinary two-valued propositions in a larger non-Boolean lattice (where they are represented by $p \mid \top$ elements) and adopting a three-valued semantics. In these settings '|' is dealt with as a three-valued truth-functional connective. Probability functions defined on the Boolean structure may be extended to the larger lattice. Since, in this setting, con-

Bruno de Finetti, Radical Probabilist
Maria Carla Galavotti (ed.)
Copyright © 2009

dition (ii) holds in general only for two-valued propositions, Lewis's problem is, at least from a formal viewpoint, solved.

Recent research (developed mainly in the '90s) on a three-valued approach to '|' was powerfully forerun by de Finetti in a paper presented at the Congress of Scientific Philosophy held in Paris in 1935 and entitled *The Logic of Probability* (de Finetti [1936] 1995).

In this paper I shall first explain de Finetti's original contribution and shall try to show the principal algebraic properties of de Finetti's logic of trievents. Second, I shall show the limits of de Finetti's contribution (as well as recent equivalent approaches) in dealing with conditional events and shall advance some proposals to overcome them. In particular, I shall introduce a new two-stages semantics (based on the new notion of *hypervaluation*) which is truth-functional in a generalized sense and allows us to solve some puzzling features of de Finetti's original approach.

2   De Finetti's logic of probability

De Finetti's reflections were stimulated by the appearance of many-valued logics to theorise probability by Łukasiewicz, Reichenbach and Mazurkiewicz. According to de Finetti, this kind of many-valued logic is epistemic in character, and affects neither the bivalent logical character of events nor the essentially subjective character of probability. In particular, adding a third truth-value meaning 'doubtful' should not "be considered as the modification which could be substituted for two-valued logic; it ought to be merely *superimposed* in considering propositions as capable, in themselves, of two values, 'true' or 'false', the distinction of 'doubtful' being only provisional and relative to O, the individual in question" (de Finetti [1936] 1995: 183).

Alongside this use of many-valued logic, de Finetti considers a second employment of it, which, in his view, may turn out to be useful in connection with conditional probability. Conditional probability is a function of a pair of propositions (or events). We may represent such a pair by a single logical entity and by a semantics with three truth-values. In other terms, if $\mathscr{A}$ is the Boolean algebra of all ordinary propositions, we may define a conditional event $(p \mid q)$, where $p, q \in \mathscr{A}$, by putting $(p \mid q) = (p, q)$. The semantics of '|' may be obtained extending any evaluation function defined over $\mathscr{A}$ $v(\varphi) = \mathscr{A} \to \{t, f\}$ to $v'(\varphi, \psi)$, where $v'(\varphi, \psi) = \mathscr{A} \times \mathscr{A} \to \{t, u, f\}$, such that:

$$v'(\varphi, \psi) = \begin{cases} t & \text{if} & v(\varphi) = t & \text{and} & v(\psi) = t \\ f & \text{if} & v(\varphi) = f & \text{and} & v(\psi) = t \\ u & \text{if} & & & v(\psi) = f \end{cases}$$

Up to this point, nothing indicates a three-valued logic. In fact, on the basis of this account, '|' is not a genuine iterable truth-functional connective and, prima facie, trievents are confined to represent pairs of ordinary propositions.

Surprisingly, de Finetti goes on to claim that "[t]he logic of trievents is a three-valued logic which can be developed in a perfect analogy to two-valued logic" (ibid.: 185). Even more surprisingly, de Finetti explains this possibility as "precisely because the trievents are only formal representations of pairs of ordinary events" (ibid.). The three-valued logic that de Finetti has in mind is reached introducing three-valued genuine connectives defined on the field of trievents. Such connectives are given by the following truth-tables:

Negation

A	¬A
t	f
u	u
f	t

Conjunction

	(A ∧ B)	A: t	u	f
	t	t	u	f
B	u	u	u	f
	f	f	f	f

Disjunction

	(A ∨ B)	A: t	u	f
	t	t	t	t
B	u	t	u	u
	f	t	u	f

Conditioning

	(A | B)	A: t	u	f
	t	t	u	f
B	u	u	u	u
	f	u	u	u

Implication

	(A → B)	A: t	u	f
	t	t	t	t
B	u	u	u	t
	f	f	u	t

Negation, Disjunction (called 'Sum' by de Finetti) and Conjunction (called 'Product' by de Finetti) coincide with Łukasiewicz's three valued corresponding truth-functions. Implication anticipates Kleene's "strong" material implication. Finally, conditioning is the novel truth-function introduced by de Finetti. It was rediscovered by several authors in relatively recent times. In particular it was rediscovered by Stephen Blamey

([1986] 2002) who had in mind the same kind of three-valued semantics that inspired de Finetti.

3   Partial logic and monotonicity

Undoubtedly, the kind of interpretation of his three-valued logic that emerges from de Finetti's work is the same that inspires Blamey's partial logic. The basic tenet of this interpretation is that the truth value 'u' means lack of a truth-value rather than a third truth value on a par with "true" and "false". It should be insisted that this is exactly de Finetti's view of his "null" (or "void") truth value. This emerges from several other sources of de Finetti's ideas. In *L'invenzione della verità* (*The Invention of Truth*), a posthumous philosophical book recently published in Italian but actually written in 1934, de Finetti wrote:

> If a distinction results in being incomplete, no harm done: it would mean that besides "true" and "false" events I would also have "null" events, or, so to speak, aborted events. As a matter of fact, it is sometimes useful to consider explicitly and intentionally from very the start such a "trievent" (especially, as will be seen later, with respect to probability theory). If, for instance, I say: "supposing that I miss the train, I shall leave by car", I am formulating a "trievent", which will be either true or false if, after missing the train, I leave by car or not, and it will be null if I do not miss the train. (de Finetti 2006: 103)

De Finetti never changed his mind on the interpretation of the third truth value. In Philosophical Lectures on Probability, another posthumous philosophical book containing a collection of oral lectures held in 1979, six years before de Finetti's death, we read:

> The introduction of the truth-value $\emptyset$ is needed in order to distinguish between reference to an event as a statement and reference to an event as a condition. Whenever the condition B is satisfied, then $A \mid B$ is either true or false (1 or 0). But unless the condition B is satisfied, one can neither say that the event $A \mid B$ is true, nor that the event $A \mid B$ is false. It is void or null, in the sense that the premise under which it is considered either true or false no longer holds. In my opinion these three cases should be treated as distinct...

It does not even mean indeterminate because an indeterminate event is an event whose truth conditions are unknown... (de Finetti 2008: 208)

De Finetti identified trievents and conditional events. The reason for this identification lies (as we shall see) in the fact that every trievent may be represented as a conditional event $A \mid B$, being A and B ordinary bivalent events. So de Finetti's identification does not entail, in principle, any loss of generality.

In his semantic analysis of partial logic, Blamey points out that the interpretation of the third value as a truth-value gap requires that logical connectives should satisfy a property called monotonicity. In this context monotonicity has to be meant with respect to the partial order between truth-values defined by $u \leq t$ and $u \leq f$ (t and f being incomparable). As Blamey says, "[a]n intuitive way to think about this is that if a formula has a value (t or f), then this value persists when any atomic gaps (u) are filled in by a value (t or f)" (Blamey [1986] 2002: 268). It is worth noticing (a) that all de Finetti's connectives are monotonic, (b) that they are functionally complete with respect to monotonicity in the sense that every monotonic truth function may be defined by the set of de Finetti's connectives, and (c) that any truth-function defined by de Finetti's connectives is monotonic. Briefly, de Finetti connectives provide a system of propositional monotonic logic. This result is straightforward in the light of Blamey's analysis. De Finetti's system contains all Kleene's strong primitive connectives with the addition of conditioning, which is the same as Blamey's transplication. Now, Blamey's system consists in turn of Kleene's strong primitive connective with the addition of another monotonic connective, called interjunction, which may be defined in terms of conditioning as $A \mid (A \to B \land B \to A)$ (Blamey [1986] 2002: 305). So de Finetti's system is equivalent to Blamey's system. Since Blamey's system defines a complete closed set of monotonic truth functions, the same holds for de Finetti's logic.

De Finetti does not explain how he found his truth tables except by following "a perfect analogy to two-valued logic". De Finetti was informed about Łukasiewicz's three-valued logic, but clearly rejected the Łukasiewicz's non monotonic material implication and bi-conditional defined by the following truth tables:

	Łukasiewicz's Implication		
$A \to_L B$	A		
	t	u	f
t	t	t	t
B u	u	t	t
f	f	u	t

	Łukasiewicz's Biconditional		
$(A \leftrightarrow_L B)$	A		
	t	u	f
t	t	u	f
B u	u	t	u
f	f	u	t

Łukasiewicz's material implication and biconditional are non-monotonic because $(A \to_L B)$ and $(A \leftrightarrow_L B)$ have the truth-value t when both A and B are null while changing B to f the truth-value changes to f. In a note in *L'invenzione della verità* de Finetti complains that Łukasiewicz's systems contain a material implication which is not well suited in representing a conditional event, but incorrectly he attributes to Łukasiewicz what actually is his (i.e. Kleene's) monotonic material implication:

> As far as I know, in mathematical logic these "trievents" are not considered, not even in the three-valued or many-valued systems of Łukasiewicz, whose theory would not be completely applicable here; the assertion "if A, then B", expressed by the formula "$A \supset B$", has a different meaning, viz. "not(A and not B)". (de Finetti 2006: 103)

We may speculate that de Finetti had not yet read Łukasiewicz's paper at the time he wrote *L'invenzione della verità*. However, since in *The Logic of Probability*, de Finetti cites Łukasiewicz's paper *Philosophical Remarks on Many-Valued Systems of Propositional Logic* ([1930] 1967), written in German (a language of which de Finetti had a near native knowledge), we should presume that he had read that paper at the time he wrote *The Logic of Probability*. Now, Łukasiewicz's paper presents at the very end of Section 6 the truth-table of his three-valued material implication (as well as of negation), so we may presume that de Finetti intentionally deviated from Łukasiewicz in constructing the truth-table for material implication, preferring what is now known as Kleene's conditional.

So we may ask why de Finetti preferred Kleene's conditional to Łukasiewicz's one. We may speculate that while de Finetti's was aimed at preserving some algebraic properties of material implication present in standard logic, Łukasiewicz's was mainly aimed, among the other things, at preserving certain basic tautologies (like $A \supset A$), that are not preserved in Kleene's system. But we may also speculate that de Finetti was guided by the "gap view" of the null truth-value, which rules out any "if-then" connective that renders a conditional assertion true when both the antecedent and the consequent are null. Monotonicity is nothing but the formal counterpart of the intuitive idea of a three-valued logic, where the third value means "null", namely absence of truth-value. Intuition of a gap interpretation of the third truth-value is a sufficiently reliable guide for finding the right truth-tables even if the abstract notion of monotonicity is not clear in one's mind.

So we may assert that in 1935 de Finetti achieved the system of Blamey's partial propositional logic, although he did not explicitly formulate the formal property that differentiates partial logic from other approaches, like Łukasiewicz's one.

It should be noticed that de Finetti's interpretation of the null truth-value is not epistemic. Trievents may be objectively either true or false or devoid of any truth-value. But this objective character is purely logical and has nothing to do with Łukasiewicz's ontological motivation about Aristotelian future contingents and determinism. Moreover, the objective attitude of de Finetti should be contrasted with the epistemological motivation of Kleene, by whom 'u' represents a kind of uncertainty about the value of a partial recursive function due to algorithmic limitations.

4   De Finetti and Bochvar

D. A. Bochvar ([1937] 1981) elaborated his three-valued logics in an attempt to face the logical antinomies of logic and set theory. Beyond Bochvar motivations, what characterizes Bochvar's system is the distinction between two classes of connectives, which he called *internal* and *external*. Bochvar's internal connectives define a logical system that coincides with the so-called Kleene's *weak* three-valued logical system, so differing from de Finetti's system. The external system contains only truth functions $f : \{u,t,f\}^k \rightarrow \{t,f\}$, so that their output is always a two-valued proposition. According to Bochvar, these connectives are considered (more or less appropriately) as "metalinguistic" in character, because they allow to say, at object-language level, that a proposition is true, false or neither true nor false.

De Finetti anticipated Bochvar's distinction and he too used, besides the monotonic connectives that we have seen, two other connectives $f : \{u,t,f\}^k \rightarrow \{t,f\}$, that are "external" in Bochvar's terminology, called respectively *Thesis* (that I shall represent by the symbol '↑', instead of de

Finetti's original 'T') and *Hypothesis* (here represented by the symbol '$\updownarrow$' instead of the original 'H'), whose truth table is the following:

De Finetti himself gives a metalinguistic interpretation of these connectives: according to him '$\uparrow X$' means "X is true", while '$\updownarrow X$' means "X is not null".

It is clear from what de Finetti says that these two connectives are

A	$\uparrow A$	$\updownarrow A$
t	t	t
u	f	f
f	f	t

meant *not* to be part of his logic of trievents. Rather, their business, according to de Finetti, is just "[t]o return from the logic of trievents to standard logic". This is provided by the following notable result (that I shall call *de Finetti's Decomposition Theorem*):

THEOREM 1 (de Finetti). (a) Every trievent may be represented by an event of the form $(A \mid B)$, where A and B are ordinary (Boolean) events; (b) given two ordinary events A and B such that B is not a tautology, $(A \mid B)$ is not an ordinary event.

Proof. (a) Let X be a trievent. Clearly both $\uparrow X$ and $\updownarrow X$ are ordinary events. Moreover, it holds: $X = (\uparrow X \mid \updownarrow X)$. This is shown by the following truth table:

(b) Let A and B be ordinary events such that B is not a tautology. B may be either false or null. In this case $(A \mid B)$ is null, which proves that $(A \mid B) \notin \mathscr{B}$. □

In Bochvar's system, '$\uparrow$' is the only primitive external connective, the other external connectives (external negation, external implication, external disjunction, external conjunction and external equivalence) being defined in terms of '$\uparrow$' and the internal connectives. De Finetti's '$\updownarrow$' (not

X	$\uparrow X$	$\updownarrow X$
t	t	t	t
u	f	u	f
f	f	f	t

null) is not explicitly present in Bochvar's system. It, however, may be

defined by '↑' and de Finetti's internal connectives, since it holds:
$\updownarrow X = \uparrow (\neg((X \to \neg X) \land (\neg X \to X)))$.

Clearly '↑' and '↕' are non-monotonic, since if X is null, both $\uparrow X$ and $\updownarrow X$ are false, whereas if they are true both $\uparrow X$ and $\updownarrow X$ are true. So the business of these connectives is just (a) to isolate the set of ordinary events, so that the Boolean algebra $\mathscr{B}$ of ordinary events may be "extracted" from the lattice $\mathscr{L}$ of trievents. Indeed, it is easy to show that $\mathscr{B} = \{X \in \mathscr{L} \mid \updownarrow X = \top\}$, where '$\top$' stands for "tautology") or, equivalently, that $\mathscr{B} = \{X \in \mathscr{L} \mid X = \uparrow X\}$, and (b) to associate every trievent one-to-one to a pair of ordinary events.

In the light of Theorem 1 we may choose to start from a Boolean algebra $\mathscr{B}$ and extend it to $\mathscr{L}$ via conditioning, since it holds: $\mathscr{L} = \{(A \mid B) : A, B \in \mathscr{B}\}$. Alternatively we may start with $\mathscr{L}$ and extract $\mathscr{B}$ in the way we have seen. This result is at the basis of de Finetti's attitude towards trievent, considering them "only a way of treating synthetically pairs of standard propositions" ([1936] 1995: 182), useful only because they express "in a clear and significant form the question concerning conditional probabilities" (ibid.). Moreover, in the light of Theorem 1, not only trievents may be constructed out of a Boolean algebra of ordinary events, but "to introduce the notion of conditional probability is to extend the definition of $\mathbf{P}(X)$ from the field of ordinary events, X, to the field of trievents" (ibid.). This (unique) extension is simply obtained by putting:

$$\mathbf{P}(E \mid H) = \frac{\mathbf{P}(E \land H)}{\mathbf{P}(H)} \tag{1}$$

provided $\mathbf{P}(H) > 0$. In this way, probability may be considered as a function of a single argument, '|' being a genuine truth-functional connective. This result provides an *ante litteram* response to Lewis's so-called triviality result. I shall return to this point later.

5   Functional completeness

Although there is strong textual evidence that de Finetti confined the logic of trievents to monotonic logic, which is a closed lattice, we may ask whether de Finetti's larger logic $\mathscr{D}$ (i.e. three-valued logic equipped with the "metalinguistic" unary operators '↑' and '↕') (a) encompasses Łukasiewicz's logic and, in such a case, (b) whether it is functionally complete (i.e., all finitary truth functions $f : \{u,t,f\}^k \to \{u,t,f\}$ can be represented as a composition of the operations of $\mathscr{D}$). To check whether de Finetti's logic encompasses Łukasiewicz's logic is a matter of deciding whether Łukasiewicz's material implication may be defined in terms of de Finetti's truth-functional connectives. The answer is positive, since it holds $(A \rightarrow_L B) = (\neg \uparrow A \lor B) \land (\neg A \lor \uparrow \neg B)$. Since the only set of connectives that occur on the right side of this identity is $\{\neg, \lor, \land, \uparrow\}$, the reduced de Finetti algebra $\mathscr{D}_R = (\{f,u,t\}, \neg, \lor, \land, \uparrow)$ encompasses Łukasiewicz's truth-functional algebra $\mathscr{L}_3 = (\{f,u,t\}, \neg, \lor, \land, \rightarrow_L)$. It is well known that $\mathscr{L}_3$ is not functionally complete.

The problem arises whether this also holds for de Finetti algebra $\mathscr{D} = (\{f,u,t\}, \neg, \lor, \land, \rightarrow, \uparrow, \updownarrow, |)$. The answer is in the negative. This result follows from a theorem proved by Słupecky in 1936 (see Malinowski 1993: 27): adding the constant truth-function $\sharp A = u$ to the stock of truth-functions in $\mathscr{L}_3$, a functionally complete algebra $\mathscr{L}_3^*$ is reached. Now, since '$\sharp$' is definable in $\mathscr{D}$ by the identity $\sharp A = A \mid (A \land \neg A)$, $\mathscr{D}$ encompasses $\mathscr{L}_3^*$, so that it turns out to be functionally complete.

6 Algebraic approach

De Finetti speaks about the notion of conditional probability as a matter of *extending* the definition of probability originally defined over a Boolean algebra to trievents. By this extension what we obtain is just a definition of a probability function over $\mathscr{D}$, not a probability function confined to the monotonic algebra of Blamey. On the other hand we may argue, from a philosophical point of view, that by considering '$\uparrow$' and '$\updownarrow$' as "metalinguistic" connectives, we are allowed to introduce them without violating the view by which the null truth-value is a truth-gap. What would be in contrast with de Finetti's view is to introduce among the *primitive* truth-functions a non-monotonic material implication (like Łukasiewicz's or Gödel's implication) as the correct generalization of two-valued material implication in the spirit of the logic of trievents. So, while much appreciating Peter Milne's *Logic of conditional assertions* – as well as its algebraic counterpart, 'de Finetti algebra' – (Milne 2004), I cannot endorse it as being in agreement with de Finetti's view of trievents.

To reconstruct de Finetti's unified views of trievents we have to consider a sentential language $\mathcal{L}$ with a denumerable set of atomic sentences equipped with the internal as well as external de Finetti connectives (namely '$\neg$', '$\lor$', '$\land$', '$\rightarrow$', '$|$', '$\uparrow$', '$\updownarrow$'). Since identity of truth conditions is a congruence relation, a quotient structure $\mathscr{L} = \langle L, \neg, \vee, \wedge, \rightarrow, |, \uparrow, \top, \bot \rangle$ results from it. Such a structure contains a Kleene algebra $\mathscr{K} = \langle L_K, \neg, \vee, \wedge, \top, \bot \rangle$, namely a bounded distributive lattice with involution satisfying De Morgan laws and the condition $p \wedge \neg p \leq p \vee \neg p$ (where '$\leq$' is the natural lattice order relation: $p \leq q$ iff $p \wedge q = p$). These properties define what Cleave (1991) calls a *normal quasi Boolean algebra*. $\mathscr{L}$, as noted before, contains a Boolean algebra of ordinary events, generated by set $\mathscr{B} = \{X \in \mathscr{L} : X = \uparrow X\}$. Moreover, it contains the constant null element '♮' definable as $\top \mid \bot$, that I shall call 'the nullity'.

7 Assertion

Asserting a sentence p equipped with the classical semantics consists of a linguistic act by which it is unconditionally claimed that p is true. Obviously, the same may not be said about sentences interpreted as trievents. Trievents may lack a truth-value, so that the unconditional assertion of their truth makes no sense. Asserting a trievent X is that linguistic act that consists of asserting that X is true provided X is not null. This is in accordance with the truth-functional identities $X = X \mid \updownarrow X = \uparrow X \mid \updownarrow X$.

The different assertability conditions between sentences expressing trievents and ordinary sentences should be taken into account in interpreting compound sentences. For example, to assert a set of ordinary sentences $\{p_1, \dots, p_n\}$ is the same as asserting their conjunction $p_1 \wedge \dots \wedge p_n$. This property cannot be extended in general to trievents. To assert separately trievents p_1 and p_2 is *not* in general the same act as asserting the compound trievent $p_1 \wedge p_2$. In fact, the proviso under which the assertion of the truth of $p_1 \wedge p_2$ is made is not that both p_1 and p_2 are not null but that $p_1 \wedge p_2$ is not null. This difference should be kept in mind to avoid fallacies and paradoxes. Most certainly, the proper use of trievents departs from the use of conditionals as made in natural language. Trievents should be used *iuxta propria principia*. To assert both p_1 and p_2 is the same as asserting the trievent $(p_1 \wedge p_2) \mid (\updownarrow p_1 \wedge \updownarrow p_2)$. This may be considered as the definition of a conjunction connective, say '$\overline{\wedge}$' whose truth table turns out to coincide with Kleene's weak conjunction:

Kleene's weak
Conjunction
$(A \bar{\wedge} B)$ A

		t	u	f
	t	t	u	f
B	u	u	u	u
	f	f	u	f

There are other ways in which sets of separate conditional assertions may be interpreted. The matter is not simple and I shall not try to elucidate this point here.

In the light of de *Finetti's Decomposition theorem* X may be represented as $E \mid H$ (where E and H are two-valued propositions) and the assertion of $E \mid H$ is that linguistic act that consists in claiming that E is true provided H is also true. The degree of the (conditional) assertability of a trievent X is precisely its probability $\mathbf{P}(X) = \mathbf{P}(X \mid \updownarrow X)$. Notice that the assertion of trievents viewed as the assertion of conditionals must be interpreted as assertion of *indicative conditionals*. Conjunctive conditionals may indeed be either true or false when the antecedent is false. Trievents with false antecedents are always neither-true-nor-false.

8 Logical consequence

A central notion in deductive logic is without question the notion of logical consequence. We may ask when a trievent q is a logical consequence of a trievent p, and, more generally, when a trievent q is a logical consequence of a set Γ of trievents. De Finetti did not address this point. However, we may try to find the best answer in the spirit of his ideas. In the literature there are several alternative definitions that extend to three-valued logics the notion of logical consequence. Łukasiewicz (and many others after him) simply required preservation of truth. This *prima facie* appears to be a natural choice. However, there are strong reasons to maintain that this requirement is inadequate. In two-valued logic, when two propositions entail each other they have the same truth-conditions and convey the same content. A reasonable requirement of adequacy for the idea of logical consequence for trievents is that this property is preserved. More formally:

(RA). $p \models q$ and $q \models p$ if and only if $p \cong q$ (where '$\cong$' stands for equality of truth-conditions).

Now, if Łukasiewicz's logical consequence definition is adopted, then RA is violated. For $q \mid p$ and $p \wedge q$ would entail each other, without having the same truth conditions. Another proposal, going back to Belnap (1973) and advocated by others, like Calabrese (1994) is the following: q is a logical consequence of p if whenever p is not false q is not false. Unfortunately, again (RA) is violated: in this case $q \mid p$ and $p \rightarrow q$ would entail each other without being truth-functionally equivalent. Moreover, as Milne (1997: 215) has pointed out, from de Finetti's standpoint by which the truth conditions of '$\mid$' are modelled on conditional bets (true when a bet is won, false when a bet is lost, null when a bet is called for), both conceptions of logical consequence are questionable. For if p is false, a bet on $q \mid p$ is called for, but a bet on $p \wedge q$ is lost and a bet on $p \rightarrow q$ is won. To satisfy RA, it is sufficient to require that both truth and non-falsehood are preserved. Any strengthening of this double requirement would have awkward side-effects. For example, it may seem reasonable to require that the null truth-value should be preserved. In this case, however, tautologies would cease to be entailed by every trievent. So the following definition of logical consequence seems to be the only suitable candidate:

Definition 1. $q \in \mathcal{L}$ is a logical consequence of $p \in Sent(\mathcal{L})$ (in symbols $p \models q$) if and only if both the following conditions are satisfied for every evaluation $v : \mathcal{L} \longrightarrow \{t,u,f\}$ of the elements of $\mathcal{L}$: (a) if $v(p) = t$ then $v(q) = t$; (b) if $v(\alpha) \in \{t,u\}$ then $v(q) \in \{t,u\}$.

In the light of the observation, made in Section 7, that the assertion of a set of sentences does not always coincide, in the context of trievents, with the assertion of their conjunction (as defined by de Finetti's truthtables), I shall avoid the extension of the preceding definition to the consequences of sets of sentences.

Definition 1 is also recommended by straightforward algebraic considerations. The main reason is that the consequence relation as defined by definition 1 induces in $\mathcal{L}$ a partial order which is isomorphic to the natural lattice order $p \leq q$ if and only if $p \wedge q = p$. This allows the characterisation of $\mathcal{L}$ as a bounded Brouwerian lattice. The truth-function that generalizes material implication of two-valued logic and provides for every $p,q \in \mathcal{L}$ the relative pseudocomplement of p relative to q is Gödel's conditional, whose truth-table is the following:

Gödel's conditional

		$(A \rightarrow_G B)$	A		
			t	u	f
		t	t	t	t
B		u	u	t	t
		f	f	f	t

9 Probability

To define a probability function on the quotient lattice $\mathcal{L}$ we may first define a probability function on the Boolean algebra $\mathcal{B}$ of ordinary events and then uniquely extend it by means of *de Finetti's Decomposition Theorem*. Indeed for every element X of $\mathcal{L}$ there are two elements E, H belonging to $\mathcal{B}$ such that $X = E \mid H$. By putting:

$$\mathbf{P}(X) = \frac{\mathbf{P}(E \wedge H)}{\mathbf{P}(H)} \text{ provided } \mathbf{P}(H) > 0,$$

the unique desired extension is obtained. Alternatively, we may try the other way round: define probability directly on $\mathcal{L}$. In such a case it also remains defined on $\mathcal{B}$ (which is contained in $\mathcal{L}$). The following axioms show how this goal may be achieved. I assume that the probability $\mathbf{P}$ is a *partially* defined function $\mathbf{P} : \mathcal{L} \longrightarrow \mathfrak{R}$. $\mathbf{P}(X)$ is not defined if (and only if) $\mathbf{P}(\updownarrow X) = 0$. In such a case there is a probability 1 that X is null, so that, believing with practical certainty that X is null, it does not make sense to assert that X is true with the proviso that it is not null. Notice that leaving undefined the probability $\mathbf{P}(X)$ of those trievents such that $\mathbf{P}(\updownarrow X) = 0$ amounts to assigning a probability value only to those trievents that, via *De Finetti's Decomposition Theorem* are representable as $X = E \mid H$ ($E, H \in \mathcal{B}$) with $\mathbf{P}(H) > 0$. The axioms that I propose are the following:

A1. If $\mathbf{P}(\updownarrow p) > 0$ then $\mathbf{P}(p) \geq 0$

A2. $\mathbf{P}(\top) = 1$

A3. If $\mathbf{P}(\updownarrow p) > 0$ then $\mathbf{P}(p) = \frac{\mathbf{P}(\uparrow p)}{\mathbf{P}(\updownarrow p)}$

A4. $\mathbf{P}(\uparrow (p \vee q)) = \mathbf{P}(\uparrow p) + \mathbf{P}(\uparrow q) - \mathbf{P}(\uparrow (p \wedge q))$

A5.   If $\mathbf{P}(\updownarrow p) > 0$ then $\mathbf{P}(\neg p) = 1 - \mathbf{P}(p)$

To show that this axiomatisation provides the same class of functions obtained by the extension method above, it suffices to show the following:

THEOREM 2. (a) If $\mathbf{P}$ is a probability function defined over $\mathscr{B}$ and (b) for every $X \in \mathscr{L}$ it holds that $\mathbf{P}(\updownarrow X) > 0$, there exist two elements $E, H \in \mathscr{B}$ such that $X = E \mid H$ and $\mathbf{P}(X) = \dfrac{\mathbf{P}(E \wedge H)}{\mathbf{P}(H)}$.

Proof. To prove (a) let p, q be any elements of $\mathscr{B}$ and let $\mathbf{P}_{\mathscr{B}}$ a probability function defined on $\mathscr{B}$. We have to show that $\mathbf{P}_{\mathscr{B}}$ satisfies all axioms A1-A5. Trivially, axioms A1 and A2 are satisfied. Since $p \in \mathscr{B}$, it holds: (i) $p = \uparrow p$ and $\updownarrow p = \top$. Since $\mathbf{P}(\top) = 1$, A3 turns out to be trivially satisfied. As far as A4 is concerned, since $\uparrow (p \vee q) = (p \vee q)$, $\uparrow p = p$, $\uparrow q = q$, and $\uparrow (p \wedge q) = (p \wedge q)$, A4 is the same on $\mathscr{B}$ as the general addition axiom, and it is obviously satisfied by $\mathbf{P}_{\mathscr{B}}$. Regarding axiom 5, since $\updownarrow p = \top$, $\mathbf{P}(\updownarrow p) = 1$ so that A5 becomes equivalent, with respect to the elements of $\mathscr{B}$, to the axiom of complement, and it is therefore satisfied. Now, A1, A2, A4, A5 provide a set of axioms for standard finite probability. Hence $\mathbf{P}_{\mathscr{B}}$ is a probability function defined over $\mathscr{B}$.

To prove (b) let $X \in \mathscr{L}$ be such that $X = E \mid H$ where $E, H \in \mathscr{B}$. By A3 it holds that

$$\mathbf{P}(X) = \mathbf{P}(E \mid H) = \frac{\mathbf{P}(\uparrow (E \mid H))}{\mathbf{P}(\updownarrow (E \mid H))} \tag{2}$$

provided $\mathbf{P}(\updownarrow (E \mid H)) > 0$. Now, $\uparrow (E \mid H) = \uparrow (E \wedge H)$. Since $E, H \in \mathscr{B}$, it holds that $\uparrow (E \wedge H) = (E \wedge H)$. Moreover, $\updownarrow (E \mid H) = (\updownarrow E \wedge \uparrow H)$ and since $E, H \in \mathscr{B}$, it holds that $\updownarrow E = \top$ and $\uparrow H = H$, so that $\updownarrow (E \mid H) = \top \wedge H = H$. Replacing, in equation (2), $\uparrow (E \mid H)$ with $E \wedge H$ and $\updownarrow (E \mid H)$ with H we obtain:

$$\mathbf{P}(X) = \frac{\mathbf{P}(E \wedge H)}{\mathbf{P}(H)} \tag{3}$$

□

This axiomatisation refers to unary non-monotonic truth-functions. The problem arises of whether a set of axioms may be found to characterise probability functions as partially defined over the monotonic sublattice of de Finetti-Blamey algebra. The answer is in the negative. In fact, as A3 shows, probability of trievents depends functionally on the probability of ordinary events. Without reference to ordinary events, no set of probability axioms are therefore possible for trievents.

10 Lewis triviality results

Without doubt, one of the main motivations for recent research on conditional assertions that led to the rediscovery of Definettian trievents comes from the philosophical impact of the so-called Lewis's triviality theorems (1976). There are several similar results all supporting the same tenet (namely that conditioning is not a connective) and essentially based on the same assumptions. The following version follows Jeffrey (2004: 15). My criticism may be easily adapted to all other existing versions. In a nutshell, the result amounts to the fact that the two following formulas:

$$((p \mid q) \mid r) = (p \mid (q \wedge r)) \tag{4}$$

$$\mathbf{P}(p) = \mathbf{P}(p \mid q)\mathbf{P}(q) + \mathbf{P}(p \mid \neg q)\mathbf{P}(\neg q) \tag{5}$$

entail:

$$\mathbf{P}(p \mid q) = \mathbf{P}(p) \tag{6}$$

which, in turn, is satisfied only in special cases of little interest. This would prove that '|' cannot be a genuine (let alone truth-functional) iterable conditional connective.

Now (4) is clearly satisfied by de Finetti's conditioning, so Lewis's result shows that (5) cannot hold in general for trievents. However, we may easily derive from axioms A1-A6 the following formula for trievents, supposing $\mathbf{P}(\updownarrow p) > 0$:

$$\mathbf{P}(p) = \frac{\mathbf{P}(\uparrow p)\mathbf{P}(\uparrow p \mid \uparrow q) + \mathbf{P}(\neg \updownarrow q)\mathbf{P}(\uparrow p \mid \neg \updownarrow q) + \mathbf{P}(\uparrow \neg q)\mathbf{P}(\uparrow p \mid \uparrow \neg q))}{\mathbf{P}(\updownarrow p)} \tag{7}$$

Clearly, (7) is a generalization of (5). For, if p and q both belong to $\mathscr{B}$, then (a) $\mathbf{P}(\updownarrow p) = \mathbf{P}(\top) = 1$, (b) $\mathbf{P}(\neg \updownarrow q) = \mathbf{P}(\bot) = 0$, (c) $\mathbf{P}(\uparrow p) = \mathbf{P}(p)$, so that (7) reduces to (5). There is no reason for requiring that (5) would be in general satisfied by trievents. In fact, (5) is just a special case (with $n = 2$) of a more general formula, called by de Finetti *conglomerative property*, valid for every *finite* partition of ordinary events $q_1, \ldots, q_n$ such that (a) $q_i \wedge q_j = \bot$ $(1 \le i < j \le n)$, and (b) $q_1 \vee \ldots \vee q_n = \top$:

$$\mathbf{P}(p) = \mathbf{P}(p \mid q_1)\mathbf{P}(q_1) + \cdots + (p \mid q_n)\mathbf{P}(q_n) \tag{8}$$

The numerator of (7) is again a special case of (8) with $n = 3$. It is perfectly natural to represent in $\mathscr{B}$ an ordinary event p by a partition of two elements (namely $\{p, \neg p\} = \{\uparrow p, \uparrow \neg p\}$) and any trievent q by a partition of three elements (namely $\{\uparrow q, \neg \updownarrow q, \uparrow \neg q\}$ that mean respectively "q is true", "q is null", "q is false"). If so, (7) is exactly the expected generalisation of (5) for trievents. The denominator in (7) is a normalisation factor. These considerations are reinforced by the following result:

THEOREM 3. Let p, q, r belonging to $\mathscr{B}$ form a partition. Let $K = \{y \in \mathscr{L} \mid \uparrow y = p, \uparrow \neg y = q, \neg \updownarrow y = r\}$. If such conditions are satisfied, then K contains one and only one element.

Proof. Let $\mathbf{V}$ be se set of all valuations $V : \mathscr{L} \longrightarrow \{t, u, f\}$. Let $y = p \mid (p \vee \neg r)$. For every valuation v in $\mathbf{V}$: (a) $v(p) = t$ iff $v(y) = t$, so that $\uparrow y = \uparrow p = p$, (b) $v(q) = t$ iff $v(y) = f$, so that $\uparrow \neg y = q$, and (c) $v(r) = t$ iff $v(y) = u$, so that $\neg \updownarrow y = r$. Therefore, $y \in K$ and $K \ne \emptyset$.

Now, suppose that there exist w_1 and w_2 both belonging to $\mathscr{L}$ such that $\uparrow w_1 = \uparrow w_2 = p$, $\uparrow \neg w_1 = \uparrow \neg w_2 = q$, and $\downarrow w_1 = \downarrow w_2 = r$. Suppose that for some v in $\mathbf{V}$ $v(w_1) \ne v(w_2)$. If $v(w_1) = t$, $v(\uparrow w_1) = t$ and $v(\uparrow w_2) = f$ against the hypothesis. Similarly, if $v(w_1) = f$, $v(\uparrow \neg w_1) = t$ and $v(\uparrow \neg w_2) = f$, against the hypothesis. Finally, if $v(w_1) = u$, $v(\neg \updownarrow w_1) = t$ and $v(\neg \updownarrow w_2) = f$ again against the hypothesis. We conclude: $w_1 = w_2$. $\square$

THEOREM 4. For every trievent $y \in \mathscr{L}$ there exists a partition of events p, q, r belonging to $\mathscr{B}$ such that $p = \uparrow y$, $q = \uparrow \neg y$, and $r = \neg \updownarrow y$.

Proof. It suffices to notice that $\uparrow y$, $\uparrow \neg y$, and $\neg \updownarrow y$ all belong to $\mathscr{B}$ and form a partition. $\square$

The conclusion we may draw is that Lewis's result was really *ante litteram* refuted by de Finetti's approach to trievents in 1935.

11 Difficulties

De Finetti's idea of trievents improves the correspondence between logic and probability, showing that absolute probability may be thought of as a generalization of sentential logic that includes partial belief and encompasses conditional probability. The latter simply becomes absolute probability of certain conditional propositions. However, from another point of view, some nice correspondences between logic and probability are lost by passing to trievents. In particular, it is well known that tautologies are, in classical theory, the only sentences such that $\mathbf{P}(p) = 1$ for every probability function. Now, for every p and every probability function $\mathbf{P}$, $\mathbf{P}(p \mid p) = 1$, but if $p \not\cong q$, then $(p \mid p) \not\cong (q \mid q)$, so that there is a multiplicity of logically non equivalent trievents of probability 1 for every probability function such that $\mathbf{P}(\updownarrow p) > 0$. The set of such trievents coincides with the set $\mathscr{F} - \mathbf{N}$, where $\mathscr{F}$ is the filter $\{x \in \mathcal{L} : \neg x \models x\}$ and $\mathbf{N} = \{x : x \cong p \mid (p \wedge \neg p)\}$.

The elements of $\mathscr{F}$ are called 'quasi-tautologies' by Bergmann (2008: 85). The elements of the dual ideal of $\mathscr{F}$, $\mathscr{I} = \{x \in \mathscr{L} : x \models \neg x\}$ are called 'quasi-contradictions'. Notice that $\mathscr{F} \cap \mathscr{I} = \mathbf{N}$, so that such sentences are, by this definition, the only elements of $\mathcal{L}$ which are at the same time quasi-tautologies and quasi-contradictions. It can be shown that every classical tautology belongs to $\mathscr{F}$ and every classical contradiction belongs to $\mathscr{I}$ (see Bergmann 2008: 86). The addition of conditioning allows us to express every quasi tautology by the form $\phi \mid \phi$. To show this, it is sufficient to notice that, given a quasi tautology ψ, it holds $\psi = \psi \mid \psi$. In fact, if ψ is true then also $\psi \mid \psi$ is true and if ψ is null also $\psi \mid \psi$ is null. The converse, however, does not hold. In particular $\mathbf{N}$ is extraneous to Kleene's logic, in the sense that the constant truth-function $f : \{t,u,f\} \longrightarrow \{u\}$ is not definable by Kleene's connectives $\neg, \vee, \wedge, \rightarrow$. Indeed it is well known that such a truth-function in not definable in Łukasiewicz's logic whereas every connective of Kleene's logic is definable in Łukasiewicz's three-valued logic.

Now the presence of $\mathscr{F}$ and $\mathscr{I}$ is a very unpleasant feature for whoever sees probability as just a generalisation of propositional logic. In classical probability, defined on Boolean algebras, the two following properties are both satisfied:

$p \leq q$ if and only if $\mathbf{P}(p) \leq \mathbf{P}(q)$ for every probability function $\mathbf{P}$.

If for every function $\mathbf{P}$, $\mathbf{P}(p) = \mathbf{P}(q)$ then $p = q$.

Now, while property 1 is preserved by the probability of trievents, since in a parallel way it holds that:

$p \leq q$ if and only if $\mathbf{P}(p) \leq \mathbf{P}(q)$ for every probability function $\mathbf{P}$ such that $\mathbf{P}(\updownarrow p) > 0$ and $\mathbf{P}(\updownarrow q) > 0$,

property 2 is not satisfied by the probability of trievents. This is substantially due to the fact that all elements that belong to $\mathscr{F}_- = \mathscr{F} - \mathbf{N}$ have probability 1 and all elements belonging to $\mathscr{I}_- = \mathscr{I} - \mathbf{N}$ have probability 0. Notice that probability is never undefined over $\mathscr{F}_-$ and $\mathscr{I}_-$, since for every $p \in \mathscr{F}_- \cup \mathscr{I}_-$, it holds that $\updownarrow p \in \mathscr{B}$. So it seems that the view of probability as a generalization of propositional logic may be extended to conditional probability by considering '|' as a genuine truth-functional connective only at the price of losing the property by which two propositions p and q are logically equivalent if for every probability function it holds that $\mathbf{P}(p) = \mathbf{P}(q)$. Progress on one side involves regress on the other.

All elements belonging to $\mathscr{F}_-$ may be represented by a sentence of the form $\psi \mid \psi$ for some ψ. Conversely, every element of $\mathcal{L}$ that is truth-functionally equivalent to a sentence of the form $\psi \mid \psi$ and does not belong to $\mathbf{N}$ belongs to $\mathscr{F}_-$. In the light of this situation, we may focus our attention on those propositions that are expressible by a sentence of the form $\psi \mid \psi$.

Moreover, we may ask in what sense, given a factual proposition p, a conditional assertion of the form "p supposed that p" may differ in content from a proposition of the same form where p is replaced with another factual proposition q. In the spirit of the interpretation of the null value as a gap, only the cases where a proposition is true or false should be taken into consideration in evaluating if a proposition is "always true" and "always false". So, if a trievent turns out to be true in all cases in which it is not null, namely in all cases in which actually *has* a truth-value, it seems natural to consider it as a tautology.

On the other hand, in calculating the probability of a trievent, only the normalized ratio between the probability that it is true and the probability that it is false is taken into consideration. The probability that the trievent in question is null is well determined by our account, but plays the role of normalizing factor. And it is just for this reason that all quasi-tautologies in $\mathscr{F}_-$ receive probability 1.

In the light of the preceding considerations, it is seems natural to provide a semantic for trievents according to which all elements in $\mathscr{F}_-$ (and, dually, all elements in $\mathscr{I}_-$) should be considered as logically equivalent (entailing each other).

An analysis from the point of view of conditional bets (that expressly guided de Finetti in finding his logic of trievents) corroborates this suggestion. Let us move from the notion (introduced by Milne 2004: 501) of *betting partial order* among sentences of $\mathcal{L}$:

Definition 2. Let p and q be sentences of $\mathcal{L}$. Then p *entails* q ($p \preceq q$) iff $\uparrow p \models \uparrow q$ and $\neg \uparrow q \models \neg \uparrow p$

The basic idea underlying this definition is that a trievent p entails a trievent q if and only given a bet B_p on p and a bet B_q on q, if B_p is won then also B_q is won and if B_q is lost then also B_p is lost. It is easy to verify that this partial order is isomorphic to the natural lattice order in $\mathcal{L}$. Let us explore whether this partial order may be refined with respect to quasi-tautologies and quasi-contradictions.

Bets on quasi-tautologies and quasi-contradictions are quite degenerate bets, being gambles without a real risk at stake. A bet on a quasi-tautology cannot be lost and a bet on a quasi-contradiction cannot be won. The difference between a bet on a quasi tautology and a bet on a full tautology lies in the fact that the latter cannot be called off, so that it is won with certainty, while the former can be called for. This, *prima facie*, justifies the difference in the partial order. However, this difference evaporates at a closer look.

First, let us observe that, in considering bets as a guide for discussing trievents with respect to logic and probability, we have to restrict our attention to *coherent* bets. Second, what defines a bet on an event is the related table of pay-offs. If every coherent bet on a trievent p has the same betting quotient of every coherent bet on a trievent q, then p and q should be considered as the same trievent. (A sentence belonging to **N** is not considered here as representing an "event" on which it makes sense to bet, since such a bet could be neither won nor lost).

Now, let us compare a coherent bet $B_\top$ on a tautology and a coherent bet $B_{p|p}$ on $p \mid p$ (where p is supposed to be neither a tautology nor a contradiction nor a nullity). The pay-offs of these coherent bets may be *necessarily*:

Outcome			Pay-offs	
p	$\top$	$p \mid p$	$B_\top$	$B_{p\mid p}$
t	Won	Won	0	0
u	Won	Called for	0	0
f	Won	Called for	0	0

so that a bet on $\top$ and on $p \mid p$ have necessarily the same betting quotient and the same pay-offs. So any coherent bet on $\top$ may be regarded objectively as an "indirect" bet on $p \mid p$. In an analogue manner it can be shown that a coherent bet on $\bot$ may be regarded as an indirect bet on $\neg(p \mid p)$ with betting quotient 0. So, according to the previous considerations, $\top$ and $p \mid p$ should be considered as the same trievent and the same may be said about $\bot$ and $\neg(p \mid p)$.

12 From supervaluations to hypervaluations

Let us now turn from the algebraic point of view to the semantical point of view. Clearly, a semantics in accordance with definition 2 cannot be purely truth-functional. The approach that I shall propose has some similarities with van Fraassen *supervaluations* (1966). Let us rehearse the latter approach to show why it cannot work in our context.

The basic idea is the following. Given a sentential language containing sentential connectives also present in the two-valued logics, first the classes of both standard three-valued and two-valued valuations are defined. A *supervaluation* is a two-stage valuation. In the first stage every formula ϕ is valuated by a three-valued valuation v in the standard way. In the second stage it is *supervaluated* by means of function v_s associated to v. The supervaluation v_s is so defined for every formula ϕ:

$v_s(\phi) = \text{t}$ if for *every* two-valued valuation w that agrees with v in the assignments of the values t and f it holds: $w(\phi) = \text{t}$;

$v_s(\phi) = \text{f}$ if for every two-valued valuation w that agrees with v in the assignments of the values t and f it holds: $w(\phi) = \text{f}$;

$v_s(\phi) = v(\phi)$ otherwise.

The original motivation of supervaluation was to save classical tautologies. In fact, every classical tautology continues to be supervaluated as true and every contradiction as false. All the rest remains unchanged. Can we resort to supervaluations in the context of trievents? The answer is in the negative. The reasons for this negative answer are the following:

1. We need to differentiate between tautologies and quasi-tautologies. This derives from the presence of **N**. Consider a sentence of the form $\phi \lor \neg\phi$ where $\phi \in \mathbf{N}$. By standard valuations $\phi \lor \neg\phi$ belongs to **N** as well. So a sentence of the form $\phi \lor \neg\phi$ cannot be considered true for every ϕ but only for those ϕ that are not nullities (i.e. do not belong to **N**). I shall call a *pre-tautology* any schema of sentence whose instances are evaluated as tautologies whenever they are not null in every event. Quasi-tautologies of de Finetti's logics of trievents that are not tautologies should be valuated as pre-tautologies.

2. Supervaluations work when all truth-functional connectives are also present in classical logic. This is not the case in the logic of trievents, due to the presence of conditioning, which is not present in classical logic. As a result, supervaluations would give to every sentence of the form $\phi \mid \phi$ the standard valuation, in spite of the fact that every such sentence is a quasi-tautology.

3. Supervaluations, appear to be reasonable with respect to Kleene's strong three-valued logic, where the set of quasi-tautology and the set of classical tautologies do coincide. In such a situation it does make sense (from a purely semantical viewpoint) to "promote" quasi-tautologies to full tautologies, leaving the rest unchanged. Once the conditioning connective is introduced, the set of quasi-tautologies no longer coincides with the set of classical tautologies.

4. Supervaluations are not compositional in character. In other terms, supervaluations have the following flaw: while a classic tautology is supervaluated as a tautology, it does not behave as a tautology in compound sentences. For example, while a sentence of the form $\phi \lor \neg\phi$ is valuated as true by every supervaluation, a sentence of the form $(\phi \lor \neg\phi) \land \psi$ is not valuated as a sentence of the form ψ by every supervaluation, as it should be expected and as it is done in classical logic.

In spite of these considerations, the basic idea of a two-stage evaluation, which introduces a modal component at the second stage, is just the kind of device we need. I shall call the kind of two-stage valuations that I shall propose *hypervaluations*.

Definition 3. Let S be the set of the sentences of L and v a valuation. The hypervaluation associated with v is the function $h_v : S \longrightarrow \{t,u,f\}$ recursively defined by the following conditions:

1. For every atomic sentence ϕ, $h_v(\phi) = v(\phi)$.
2. If $\phi = \neg\psi$ then
   a. $h_v(\phi) = t$ if $h_v(\psi) = f$;
   b. $h_v(\phi) = f$ if $h_v(\psi) = t$;
   c. $h_v(\phi) = u$ otherwise.
3. If $\phi = (\chi \vee \psi)$ then
   a. $h_v(\phi) = t$ if at least one of the following conditions are satisfied:
      i. $h_v(\chi) = t$;
      ii. $h_v(\psi) = t$;
      iii. All the following conditions are satisfied:
         01. for no valuation w, $h_w(\chi) = f$ and $h_w(\psi) = f$;
         02. there exists a valuation w' such that either $h_{w'}(\chi) = t$ or $h_{w'}(\psi) = t$.
   b. $h_v(\phi) = f$ if at least one of the following conditions are satisfied:
      i. $h_v(\chi) = f$ and $h_v(\psi) = f$;
      ii. All the following conditions are satisfied:
         01. for every valuation w: $h_w(\chi) \in \{u,f\}$ and $h_w(\psi) \in \{u,f\}$;
         02. there exists a valuation w' such that $h_{w'}(\chi) = f$ and $h_{w'}(\psi) = f$
   c. $h_v(\phi) = u$ otherwise.
4. If $\phi = (\chi \wedge \psi)$ then
   a. $h_v(\phi) = t$ if at least one of the following conditions are satisfied:
      i. $h_v(\chi) = t$ and $h_v(\psi) = t$;
      ii. All the following conditions are satisfied:
         01. for every valuation w: $h_w(\chi) \in \{t,u\}$ and $h_w(\psi) \in \{t,u\}$;
         02. there exists a valuation w' such that $h_{w'}(\chi) = t$ and $h_{w'}(\psi) = t$.

    b. $h_v(\phi) = f$ if at least one of the following conditions are satisfied:

        i.   Either $h_v(\chi) = f$ or $h_v(\psi) = f$;

        ii. All the following conditions are satisfied:

            01. for no valuation w: $h_w(\chi) = t$ and $h_w(\psi) = t$;

            02. there exists a valuation w' such that either $h_{w'}(\chi) = f$ or $h_{w'}(\psi) = f$

    c. $h_v(\phi) = u$ otherwise.

5. If $\phi = (\chi \rightarrow \psi)$ then $h_v(\phi) = h_v(\neg\chi \lor \psi)$

6. If $\phi = (\chi \mid \psi)$ then

    a. $h_v(\phi) = t$ if at least one of the following conditions are satisfied:

        i. $h_v(\chi) = t$ and $h_v(\psi) = t$;

        ii. All the following conditions are satisfied:

            01. for every valuation w such that $h_w(\psi) = t$, $h_w(\chi) \in \{t, u\}$

            02. there is a valuation w' such that $h_w(\chi) = t$ and $h_w(\psi) = t$

    b. $h_v(\phi) = f$ if at least one of the following conditions are satisfied:

        i. $h_v(\chi) = f$ and $h_v(\psi) = t$;

        ii. All the following conditions are satisfied:

            01. for every valuation w such that $h_w(\psi) = t$, $h_w(\chi) \in \{f, u\}$

            02. there is a valuation w' such that $h_w(\chi) = f$ and $h_w(\psi) = t$

    c. $h_v(\phi) = u$ otherwise.

7. If $\phi = \uparrow \psi$ then

    a. $h_v(\phi) = t$ if $h_v(\psi) = t$

    b. $h_v(\phi) = f$ otherwise.

8. If $\phi = \updownarrow \psi$ then

    a. $h_v(\phi) = t$ if either $h_v(\psi) = t$ or $h_v(\psi) = f$;

    b. $h_v(\phi) = f$ otherwise.

The *compositional* character of hypervaluations should be stressed in a comparison with supervaluations. This allows us to consider the logical employment of non-null quasi-tautologies as full tautologies. In turn, this also permits us to define consistently the semantic notions of logical consequence and logical equivalence.

Definition 4. The sentences ϕ and ψ of $\mathcal{L}$ are *semantically equivalent* to each other ($\phi \approx \psi$) iff for every valuation v: $h_v(\phi) = h_v(\psi)$.

Definition 5. The sentence ϕ of $\mathcal{L}$ is a *pre-tautology* iff for every valuation w: $h_w(\phi) \in \{t,u\}$ while there exists a valuation w' such that $h_{w'}(\phi) = t$.

Definition 6. The sentence ϕ of $\mathcal{L}$ is a *pre-contradiction* iff for every valuation w: $h_w(\phi) \in \{f,u\}$ while there exists a valuation w' such that $h_{w'}(\phi) = f$.

Definition 7. The sentence ϕ of $\mathcal{L}$ is *factual* iff there exist valuations v, w such that $h_v(\phi) = t$, $h_w(\phi) = f$

Definition 8. The sentence ϕ of $\mathcal{L}$ is *void* iff for every valuation w
$$h_w(\phi) = u.$$

Definition 3 introduces a modal (intensional) component (i.e. a dependence from the whole set of hypervaluations) so modifying the truth-conditions of the monotonic binary connectives (maintaining the purely truth-functional character of all the unary operators).

The question arises whether $\mathcal{L}$ gives rise to a lattice with respect to '$\vee$' and '$\wedge$'. The answer is in the positive as is proved by the following results:

THEOREM 5. For every ϕ, (a) if ϕ is pre-tautology (pre-contradiction) then ϕ is a tautology, i.e. for every valuation w, $h_w(\phi) = t$ (contradiction, i.e. for every valuation w, $h_w(\psi) = f$).

Proof. We consider separately an exhaustive list of mutually exclusive cases:
1. If ϕ is atomic the thesis is vacuously true because ϕ cannot be either a pre-tautology or a pre-contradiction;

2. If $\phi = \neg\psi$ and ψ is pre-contradiction (pre-tautology) then by definition 3, clause a (i), (clause b (i)) for every valuation v $h_v(\psi) = f$ $(h_v(\psi) = t)$ so that ψ is a contradiction (tautology) and ϕ is a tautology (contradiction);

3. If $\phi = \chi \vee \psi$ then either χ or ψ is a pre-tautology (both χ and ψ are pre-contradictions) and by definition 3, clauses a (i) and (ii) (clause b (iv)), for every valuation v $h_v(\phi) = t$ $(h_v(\phi) = f)$, so that ϕ is a tautology (contradiction);

4. If $\phi = \chi \wedge \psi$ then both χ and ψ are pre-tautologies (either χ or ψ is a pre-contradiction) and by definition 3, clause a (i) (clauses (i) and (ii)), for every valuation v $h_v(\phi) = t$ $(h_v(\phi) = f)$, so that ϕ is a tautology (contradiction);

5. If $\phi = \chi \rightarrow \psi$ either χ is a pre-contradiction (pre-tautology) or (and) ψ is a pre-tautology (pre-contradiction). Since, by definition 3, clause 5, for every v $h_v(\phi) = h_v(\neg\chi \vee \psi)$, and by clause 2 for every v $h_v(\neg\chi) = t$ $(h_v(\neg\chi) = f)$, by clause 3a (iii) $h_v(\phi) = t$ (by clause 3b (iv) $h_v(\phi) = f$) so that ϕ is a tautology (contradiction);

6. If $\phi = \chi \mid \psi$, then also $\psi \rightarrow \chi$ is a pre-tautology ($\chi \wedge \psi$ is a pre-contradiction), so that, as proved at point 5 (4) above ϕ is a tautology (a contradiction);

7. If $\phi = \uparrow \psi$ then ψ is a pre-tautology (pre-contradiction). By definition 3.7, clause a (ii) (clause b) for every valuation v $h_v(\phi) = t$ $(h_v(\phi) = f)$, so that ϕ is a tautology (contradiction);

8. If $\phi = \updownarrow \psi$ then ψ is either a pre-tautology or a pre-contradiction (void). By definition 3.8, clause a (ii) and (iii) (clause b) for every valuation v $h_v(\phi) = t$ $(h_v(\phi) = f)$, so that ϕ is a tautology (contradiction). $\square$

Theorem 5 shows that every sentence ϕ is either (a) factual, or (b) a contradiction, or (c) a tautology, or (d) void. So our semantics allows us to get rid of pre-tautologies and pre-contradictions, converting them compositionally into tautologies and contradictions respectively. However two main differences with classical logic should be noticed.

First, no unrestricted substitution rule holds. As a result, tautologies and contradictions may not be expressed by unconditioned *schemas*. For example, the schema $\phi \vee \neg\phi$ does *not* in itself represent a class of "valid" sentences in the same sense as the same schema represents a class of quasi-tautologies in standard three-valued logic. The point is that not all instances of that schema are tautologies. More exactly, a sentence of the form $\phi \vee \neg\phi$ is not valid if (and only if) ϕ is a nullity. We may again use schemas with appropriate provisos: "if ϕ is not a nullity then $\phi \vee \neg\phi$ is a tautology". The clause "if ϕ is not a nullity" cannot in general be immediately decided by inspecting ϕ. What, however is essential is that this property is algorithmically decidable, and this turns out to be the case[1].

The second difference with the traditional approach lies in the fact that the truth-conditions of sentences are not given by simple truth-tables. A modal component (i.e. a reference to the set of all valuations) is essentially present in definition 3. In spite of this, the truth-value of a sentence depends only on the valuations of the atomic sentences that occur in it. This is shown by the following result:

THEOREM 6. Let ϕ be a sentence of $\mathcal{L}$. Let $\psi_1, \ldots, \psi_n$ be all the atomic sentences occurring in ϕ. If two valuations v, v' are such that for every i $(1 \leq i \leq n)$ $v(\psi_i) = v'(\psi_i)$, then $h_v(\phi) = h_{v'}(\phi)$.

Proof. Let us proceed by course-of-values induction on the number n of connectives occurring in ϕ. If $n = 0$ then ϕ is atomic, so that, by Definition 3, clause 1, $h_v(\phi) = v(\phi) = v'(\phi) = h_{v'}(\phi)$. Now suppose that $n = m + 1$ and that for every $k \leq m$ the theorem is true. We have the following exhaustive list of cases:
1. $\phi = \wr\psi$ (where '$\wr$' stands for any unary connective). By inductive hypothesis $h_v(\psi) = h_{v'}(\psi)$ and by Definition 3, clause 2 and clauses 7-8, in every case $h_v(\phi) = h_{v'}(\phi)$;
2. $\phi = \chi \bullet \psi$ (where '$\bullet$' stands for any binary connective). By inductive hypothesis $h_v(\chi) = h_{v'}(\chi)$ and $h_v(\psi) = h_{v'}(\psi)$, so that by Definition 3, clauses 3-6, in every case $h_v(\phi) = h_{v'}(\phi)$. $\square$

In the light of Theorem 6 we may ask whether, in a generalised sense, our logic, in spite of its semi-modal character may still be considered as truth-functional. A $n - $ ary sentential connective $\circledast$ is truth-functional in the classical sense (or strict sense) if the truth-value of any sentence $\circledast(\phi_1, \ldots, \phi_n)$ is a function of the truth-values of the sentences $\phi_1, \ldots, \phi_n$. Clearly, the binary connectives of $\mathcal{L}$, according to definition 3, are not truth-functional in this sense. However, we may consider the following more general definition of truth-functionality:

---

[1] As we shall see later, it is possible to decide algorithmically if a sentence is void by our mutating truth tables. This property would not be preserved if the present theory were extended to quantification theory.

Definition 9. A *n* − ary sentential connective ⊛ is *truth-functional in the generalized sense* iff the truth-value of any sentence $\psi = \circledast(\phi_1,\ldots,\phi_n)$ is a function of the truth-values of the atomic sentences $p_1,\ldots,p_m$ occurring in ψ.

Clearly the semantics of the binary connectives as characterised by definition 3 is truth-functional in the generalised sense.[2] This allows us to employ a truth-table technique in determining the truth-value of a molecular sentence for every assignment of truth-values to the atomic sentences that occur in it. I shall call this technique the 'mutant truth-tables procedure'. The basic idea is very simple. In developing the original truth-table algorithm, every computed column is checked for pre-tautology or pre-contradiction and it is immediately converted into a tautology or contradiction respectively *before the process continues*.

Consider, for example, the sentence '$(p \mid p) \wedge (\neg p \mid \neg p)$'. The standard truth-table for this sentence is the following (where the symbol '▼' indicates the pre-tautological columns):

(p	|	p)	∧	(¬	p	|	¬	p)	p
	▼				▼				
t	t	t	u	f	t	u	f	t	t
u	u	u	u	u	u	u	u	u	u
f	u	f	u	t	f	t	t	f	f

The mutant truth-tables algorithm is, in contrast, the following:

(p	|	p)	∧	(¬	p	|	¬	p)	p
	▼				▼				
t	t	t	t	f	t	~~u~~t	f	t	t
u	~~u~~t	u	t	u	u	~~u~~t	u	u	u
f	~~u~~t	f	t	t	f	t	t	f	f

So according to de Finetti's original semantics a sentence like "supposed that it is raining it is raining and supposed that it is not raining it is not raining" is uniformly null, while according to the semantics here proposed (and to common use of natural language) it is just a tautology.

---

[2] By the way, we notice that the generalized sense was what Wittgenstein actually defined in *Tractatus logico-philosophicus*, 5: "Propositions are truth-functions of elementary propositions", in spite of the fact that he considered only truth-functions in the strict sense.

13 Back to algebra

Let S be the set of the sentences of $\mathcal{L}$. Let $S/\approx$ be the set of equivalence classes induced by the relation $\approx$. Let us introduce in $S/\approx$ the operations $\cup, \cap$, by putting: $|\phi| \cup |\psi| = |\phi \vee \psi|$ and $|\phi| \cap |\psi| = |\phi \wedge \psi|$. Let $\mathcal{A} = \langle S/\approx, \cup, \cap \rangle$ the resulting quotient algebra. The first step in our attempt to prove the adequacy of our semantics is to prove that $\mathcal{A}$ is a lattice. Before going on to prove this I shall prove some auxiliary result.

THEOREM 7. (a) If ϕ is void and ψ is either void or a contradiction then $\phi \vee \psi$ is void; (b) if ϕ is void and ψ is neither void nor a contradiction then $\phi \vee \psi$ is a tautology.

THEOREM 8. (a) If ϕ is void and ψ is either void or a tautology then $\phi \wedge \psi$ is void; (b) if ϕ is void and ψ is neither void nor a tautology then $\phi \wedge \psi$ is a contradiction.

Theorems 7 and 8 show that when a void sentence (i.e. an uniformly null sentence) is involved, the semantics of conjunction and disjunction deviates from standard logic. If truth-values are expressed numerically (for example by 0, ½, 1 standing for false, null and true respectively), the standard properties by which for every valuation $v(\phi \vee \psi) = \max(v(\phi), v(\psi))$ and $v(\phi \wedge \psi) = \min(v(\phi), v(\psi))$ do not extend to hypervaluations. This happens every time a sentence of the form $(\phi \vee \psi)$ is a pre-tautology by standard valuation and is changed in a tautology, or a sentence of the form $(\phi \wedge \psi)$ is a pre-contradiction by standard valuation and is changed in a contradiction. However, as it is clear, the following inequalities are retained:

$$h_v(\phi \vee \psi) \geq \max(h_v(\phi), h_v(\psi)) \text{ and}$$
$$h_v(\phi \wedge \psi) \leq \min(h_v(\phi), h_v(\psi)).$$

The following result shows that the relation of logical equivalence defined in terms of hypervaluations is a congruence relation.

THEOREM 9. The relation $\approx$ is a congruence.
Proof. Clearly $\approx$ is reflexive, symmetric and transitive. It remains to prove:
a. If $\phi \approx \psi$ then (a.i) $\neg\phi \approx \neg\psi$, (a.2) $\uparrow \phi \approx \uparrow \psi$, (a.3) $\updownarrow \phi \approx \updownarrow \psi$.

By definition 3, $h_v(\neg\varphi)$, $h_v(\uparrow\varphi)$, and $h_v(\updownarrow\varphi)$ are, for every v, determined by $h_v(\varphi)$, so that if (as $\phi\approx\psi$ implies) for every v $h_v(\phi) = h_v(\psi)$ it follows (a.1) $\neg\phi\approx\neg\psi$, (a.2) $\uparrow\phi\approx\uparrow\psi$, and (a.3) $\updownarrow\phi\approx\updownarrow\psi$.

b. If $\phi_1\approx\psi_1$ and $\phi_2\approx\psi_2$ then (b.1) $\phi_1\vee\phi_2\approx\psi_1\vee\psi_2$, (b.2) $\phi_1\wedge\phi_2\approx\psi_1\wedge\psi_2$, (b.3) $\phi_1\rightarrow\phi_2\approx\psi_1\rightarrow\psi_2$, (b.4) $\phi_1\mid\phi_2\approx\psi_1\mid\psi_2$. It is readily seen that, by definition 3, if for every w $h_w(\phi_1) = h_w(\phi_2)$ and $h_w(\psi_1) = h_w(\psi_2)$, for any valuation v $h_v(\phi_1\vee\phi_2) = h_v(\psi_1\vee\psi_2)$, $h_v(\phi_1\wedge\phi_2) = h_v(\psi_1\wedge\psi_2)$, $h_v(\phi_1\rightarrow\phi_2) = h_v(\psi_1\rightarrow\psi_2)$, $h_v(\phi_1\mid\phi_2) = h_v(\psi_1\mid\psi_2)$, so that (b.1) $\phi_1\vee\phi_2\approx\psi_1\vee\psi_2$, (b.2) $\phi_1\wedge\phi_2\approx\psi_1\wedge\psi_2$, (b.3) $\phi_1\rightarrow\phi_2\approx\psi_1\rightarrow\psi_2$, (b.4) $\phi_1\mid\phi_2\approx\psi_1\mid\psi_2$.
□

Theorem 9 allows us to define the quotient structure $\mathcal{A} = \mathcal{S}/\approx$. The crucial problem arises of whether $\mathcal{A}$ is a lattice. Given the nonstandard character of hypervaluation semantics this is not a trivial problem. The positive answer is provided by the following result.

THEOREM 10. $\langle\mathcal{A}; \cup, \cap\rangle$ is a bounded lattice.

Proof. First, observe that $\mathcal{A}$ is a nonempty set and that $\cup, \cap$ are binary operations on $\mathcal{A}$. We have to prove: (i) $|\phi|\cup|\psi| = |\psi|\cup|\phi|$; (ii) $|\phi|\cap|\psi| = |\psi|\cap|\phi|$; (iii) $(|\phi|\cup|\psi|)\cup|\chi| = |\phi|\cup(|\psi|\cup|\chi|)$; (iv) $(|\phi|\cap|\psi|)\cap|\chi| = |\phi|\cap(|\psi|\cap|\chi|)$; (v) $|\phi|\cup(|\phi|)\cap|\psi|) = |\phi|$; (vi) $|\phi|\cap(|\phi|)\cup|\psi|) = |\phi|$; (vii) there exists $\phi\in\mathcal{S}$ such that for every $\psi\in\mathcal{S}$, $|\phi|\cup|\psi| = |\psi|$; (viii) there exists $\phi\in\mathcal{S}$ such that for every $\psi\in\mathcal{S}$, $|\phi|\cap|\psi| = |\psi|$.

i. Let $\phi, \psi\in\mathcal{S}$. From definition 3 there is a perfect symmetry between the truth-conditions of $\phi\vee\psi$ and the truth-conditions of $\psi\vee\phi$, so that for every valuation v it holds that $h_v(\phi\vee\psi) = h_v(\psi\vee\phi)$. Therefore, $\phi\vee\psi\approx\psi\vee\phi$, so that $|\phi\vee\psi| = |\psi\vee\phi|$ and $|\phi|\cup|\psi| = |\psi|\cup|\phi|$.

ii. Let $\phi, \psi\in\mathcal{S}$. From definition 3 there is a perfect symmetry between the truth-conditions of $\phi\wedge\psi$ and the truth-conditions of $\psi\wedge\phi$, so that for every valuation v it holds that $h_v(\phi\wedge\psi) = h_v(\psi\wedge\phi)$. Therefore, $\phi\wedge\psi\approx\psi\wedge\phi$, so that $|\phi\wedge\psi| = |\psi\wedge\phi|$ and $|\phi|\cap|\psi| = |\psi|\cap|\phi|$.

iii. Let $\phi, \psi, \chi\in\mathcal{S}$. We have first to prove that for every valuation v it holds that $h_v((\phi\vee\psi)\vee\chi)) = h_v(\phi\vee(\psi\vee\chi))$. Let $g : \{t,u,f\} \longrightarrow \{0, \frac{1}{2}, 1\}$ such that $g(f) = 0$, $g(u) = \frac{1}{2}$, $g(t) = 1$ and let

$f_v : \mathcal{S} \longrightarrow \{0, \frac{1}{2}, 1\} = g \circ h_v$. Let us distinguish the following exhaustive list of cases:

a. $f_v(\phi \vee \psi) = \max(f_v(\phi), f_v(\psi)), f_v(\phi \vee \chi) = \max(f_v(\phi), f_v(\chi))$
   and $f_v(\psi \vee \chi) = \max(f_v(\psi), f_v(\chi))$

b. $f_v(\phi \vee \psi) = \max(f_v(\phi), f_v(\psi)), f_v(\phi \vee \chi) = \max(f_v(\phi), f_v(\chi))$
   and $f_v(\psi \vee \chi) = 1$

c. $f_v(\phi \vee \psi) = \max(f_v(\phi), f_v(\psi)), f_v(\phi \vee \chi) = 1$, and
   $f_v(\psi \vee \chi) = \max(f_v(\psi), f_v(\chi))$

d. $f_v(\phi \vee \psi) = \max(f_v(\phi), f_v(\psi)), f_v(\phi \vee \chi) = 1$, and
   $f_v(\psi \vee \chi) = 1$

e. $f_v(\phi \vee \psi) = 1, f_v(\psi)), f_v(\phi \vee \chi) = \max(f_v(\phi), f_v(\chi))$ and
   $f_v(\psi \vee \chi) = \max(f_v(\psi), f_v(\chi))$

f. $f_v(\phi \vee \psi) = 1, f_v(\phi \vee \chi) = \max(f_v(\phi), f_v(\chi))$, and
   $f_v(\psi \vee \chi) = 1$

g. $f_v(\phi \vee \psi) = 1, f_v(\phi \vee \chi) = 1$, and
   $f_v(\psi \vee \chi) = \max(f_v(\psi), f_v(\chi))$

h. $f_v(\phi \vee \psi) = 1, f_v(\phi \vee \chi) = 1$, and $f_v(\psi \vee \chi) = 1$.

In the case (a), (iii) is obviously satisfied by the well-known properties of maximum.

In the case (b), $f_v(\phi \vee (\psi \vee \chi)) = 1$ and since it holds that
$f_v(\phi \vee \psi) \vee \chi)) \geq f_v(\perp \vee \psi) \vee \chi)) = f_v(\psi \vee \chi)$,
(iii) is satisfied too.

In the case (c), since $f_v(\phi \vee \psi) \geq f_v(\perp \vee \phi) = f_v(\phi)$ and $f_v(\phi \vee \chi)) = 1$,
$h_v((\phi \vee \psi) \vee \chi)) = 1$.

On the other hand $1 = f_v(\phi \vee \chi)) \leq f_v(\phi \vee (\psi \vee \chi)) = 1$.

In the case (d) $1 = f_v(\psi \vee \chi) \leq f_v(\phi \vee (\psi \vee \chi)) = 1$ and
$1 = f_v(\phi \vee \chi) \leq f_v(\phi \vee (\psi \vee \chi)) = 1$

In the case (e) $h_v((\phi \vee \psi) \vee \chi)) = h_v(\top \vee \chi) = 1$ and
$1 = f_v(\phi \vee \psi) \leq f_v(\phi \vee (\psi \vee \chi)) = 1$.

In the case (f) $1 = f_v(\phi \vee \psi) \leq f_v((\phi \vee \psi) \vee \chi) = 1$ and
$1 = f_v(\psi \vee \chi) \leq f_v(\phi \vee (\psi \vee \chi)) = 1$

In cases (g) and (h), $1 = f_v(\phi \vee \psi) \leq h_v((\phi \vee \psi) \vee \chi) = 1$ and
$1 = h_v(\phi \vee \chi) \leq h_v(\phi \vee (\psi \vee \chi)) = 1$

From $h_v((\phi \vee \psi) \vee \chi)) = h_v(\phi \vee (\psi \vee \chi))$ for every v and every ϕ, ψ, χ
it follows $|((\phi \vee \psi) \vee \chi))| = |(\phi \vee (\psi \vee \chi))|$,
so that $((\phi \vee \psi) \vee \chi)) \approx (\phi \vee (\psi \vee \chi))$.

iv.   Let $\phi, \psi, \chi \in \mathcal{S}$. We have first to prove that for every valuation v it
holds that $h_v((\phi \wedge \psi) \wedge \chi)) = h_v(\phi \wedge (\psi \wedge \chi))$. Let

$g: \{t,u,f\} \longrightarrow \{0,\frac{1}{2},1\}$ and $f_v : \mathcal{S} \longrightarrow \{0,\frac{1}{2},1\}$ as in (iii). Let us distinguish the following exhaustive list of cases:

a. $f_v(\phi \wedge \psi) = \min(f_v(\phi), f_v(\psi)), f_v(\phi \wedge \chi) = \min(f_v(\phi), f_v(\chi))$
and $f_v(\psi \wedge \chi) = \min(f_v(\psi), f_v(\chi))$

b. $f_v(\phi \wedge \psi) = \min(f_v(\phi), f_v(\psi)), f_v(\phi \wedge \chi) = \min(f_v(\phi), f_v(\chi))$
and $f_v(\psi \wedge \chi) = 0$

c. $f_v(\phi \wedge \psi) = \min(f_v(\phi), f_v(\psi)), f_v(\phi \wedge \chi) = 0$, and
$f_v(\psi \wedge \chi) = \min(f_v(\psi), f_v(\chi))$

d. $f_v(\phi \wedge \psi) = \min(f_v(\phi), f_v(\psi)), f_v(\phi \wedge \chi) = 0$, and
$f_v(\psi \wedge \chi) = 0$

e. $f_v(\phi \wedge \psi) = 0, f_v(\psi)), f_v(\phi \wedge \chi) = \min(f_v(\phi), f_v(\chi))$ and
$f_v(\psi \wedge \chi) = \min(f_v(\psi), f_v(\chi))$

f. $f_v(\phi \wedge \psi) = 0, f_v(\phi \wedge \chi) = \min(f_v(\phi), f_v(\chi))$, and
$f_v(\psi \wedge \chi)| = 0$

g. $f_v(\phi \wedge \psi) = 0, f_v(\phi \wedge \chi) = 0$, and
$f_v(\psi \wedge \chi) = \min(f_v(\psi), f_v(\chi))$

h. $f_v(\phi \wedge \psi) = 0, f_v(\phi \wedge \chi) = 0$, and $f_v(\psi \wedge \chi) = 0$.

In the case (a) (iv) is obviously satisfied by the well-known properties of minimum.

In the case (b), $f_v(\phi \wedge (\psi \wedge \chi)) = 0$ and since
$f_v(\phi \wedge \psi) \wedge \chi)) \leq f_v(\top \wedge \psi) \wedge \chi)) = f_v(\psi \wedge \chi)$, (iv) is satisfied too.

In the case (c), since $f_v(\phi \wedge \psi) \leq f_v(\top \wedge \phi) = f_v(\phi)$ and $f_v(\phi \wedge \chi)) = 0$,
$h_v((\phi \wedge \psi) \wedge \chi)) = 0$. On the other hand
$0 = f_v(\phi \wedge \chi)) \geq f_v(\phi \wedge (\psi \wedge \chi)) = 0$.

In the case (d) $0 = f_v(\psi \wedge \chi) \geq f_v(\phi \wedge (\psi \wedge \chi)) = 0$ and
$0 = f_v(\phi \wedge \chi) \geq f_v(\phi \wedge (\psi \wedge \chi)) = 0$

In the case (e) $h_v((\phi \wedge \psi) \wedge \chi)) = h_v(\bot \wedge \chi) = 0$ and
$0 = f_v(\phi \wedge \psi) \geq f_v(\phi \wedge (\psi \wedge \chi)) = 0$.

In the case (f) $0 = f_v(\phi \wedge \psi) \geq f_v((\phi \wedge \psi) \wedge \chi) = 0$ and
$0 = f_v(\psi \wedge \chi) \geq f_v(\phi \wedge (\psi \wedge \chi)) = 0$.

In cases (g) and (h), $0 = f_v(\phi \wedge \psi) \geq h_v((\phi \wedge \psi) \wedge \chi) = 0$ and
$0 = h_v(\phi \wedge \chi) \geq h_v(\phi \wedge (\psi \wedge \chi)) = 0$.

From $h_v((\phi \wedge \psi) \wedge \chi)) = h_v(\phi \wedge (\psi \wedge \chi))$ for every v and every ϕ, ψ, χ it follows that $|((\phi \wedge \psi) \wedge \chi))| = |(\phi \wedge (\psi \wedge \chi))|$, so that
$((\phi \wedge \psi) \wedge \chi)) \approx (\phi \wedge (\psi \wedge \chi))$.

v. Let $\phi, \psi \in \mathcal{S}$. We have first to prove that for every valuation v it holds that $h_v(\phi \vee (\phi \wedge \psi)) = h_v(\phi)$. Let $g: \{t,u,f\} \longrightarrow \{0,\frac{1}{2},1\}$ and

$f_v : \mathcal{S} \longrightarrow \{0, \frac{1}{2}, 1\}$ as in (iii). Let us distinguish the following exhaustive list of cases:

    a. $f_v(\phi \wedge \psi) = \min(f_v(\phi), f_v(\psi))$ and
        $f_v(\phi \vee (\phi \wedge \psi)) = \max(f_v(\phi), f_v(\phi \wedge \psi))$

    b. $f_v(\phi \wedge \psi) = \min(f_v(\phi), f_v(\psi))$ and $f_v(\phi \vee (\phi \wedge \psi)) = 1$

    c. $f_v(\phi \wedge \psi) = 0$ and $f_v(\phi \vee (\phi \wedge \psi)) = \max(f_v(\phi), f_v(\phi \wedge \psi))$

    d. $f_v(\phi \wedge \psi) = 0$ and $f_v(\phi \vee (\phi \wedge \psi)) = 1$

In case (a) the thesis is obvious by standard properties of maximum and minimum.

In case (b) if $\min(f_v(\phi), f_v(\psi)) = f_v(\phi)$, then $1 = f_v(\phi \vee \phi) = f_v(\phi)$, so that $h_v(\phi \vee (\phi \wedge \psi)) = h_v(\phi)$. If $\min(f_v(\phi), f_v(\psi)) = f_v(\psi)$ then $f_v(\phi) \geq f_v(\psi)$ so that if $f_v(\phi) \in \{0, \frac{1}{2}\}$ also $f_v(\psi) \in \{0, \frac{1}{2}\}$ and by definition 3 in no case $f_v(\phi \vee (\phi \wedge \psi)) = f_v(\phi \vee \psi) = 1$. So $f_v(\phi) \in \{\frac{1}{2}, 1\}$. If $f_v(\phi) = \frac{1}{2}$ then $f_v(\psi) = \frac{1}{2}$ but then in every case by definition 3 $f_v(\phi \vee (\phi \wedge \psi)) = \frac{1}{2}$. It follows that $f_v(\phi) = 1$ in which case $f_v(\phi \vee (\phi \wedge \psi)) = 1 = f_v(\phi)$.

In case (c) the thesis is trivially satisfied.

In case (d) $f_v(\phi \vee (\phi \wedge \psi)) = 1$ may be satisfied only if $f_v(\phi) = 1$.

vi.   Let $\phi, \psi \in \mathcal{S}$. We have first to prove that for every valuation v it holds that $h_v(\phi \wedge (\phi \vee \psi)) = h_v(\phi)$. Let $g : \{t, u, f\} \longrightarrow \{0, \frac{1}{2}, 1\}$ and $f_v : \mathcal{S} \longrightarrow \{0, \frac{1}{2}, 1\}$ as in (iii). Let us distinguish the following exhaustive list of cases:

    a. $f_v(\phi \vee \psi) = \max(f_v(\phi), f_v(\psi))$ and
        $f_v(\phi \wedge (\phi \vee \psi)) = \min(f_v(\phi), f_v(\phi \vee \psi))$

    b. $f_v(\phi \vee \psi) = \max(f_v(\phi), f_v(\psi))$ and $f_v(\phi \wedge (\phi \vee \psi)) = 0$

    c. $f_v(\phi \vee \psi) = 1$ and $f_v(\phi \wedge (\phi \vee \psi)) = \min(f_v(\phi), f_v(\phi \vee \psi))$

    d. $f_v(\phi \vee \psi) = 1$ and $f_v(\phi \wedge (\phi \vee \psi)) = 0$

In case (a) the thesis is obvious by standard properties of maximum and minimum.

In case (b) if $\max(f_v(\phi), f_v(\psi)) = f_v(\phi)$, then $0 = f_v(\phi \wedge \phi) = f_v(\phi)$, so that $h_v(\phi \vee (\phi \wedge \psi)) = h_v(\phi)$. If $\max(f_v(\phi), f_v(\psi)) = f_v(\psi)$ then $f_v(\phi) \leq f_v(\psi)$ so that if $f_v(\phi) \in \{\frac{1}{2}, 1\}$ also $f_v(\psi) \in \{\frac{1}{2}, 1\}$ and by definition 3 in no case $f_v(\phi \wedge (\phi \vee \psi)) = f_v(\phi \wedge \psi) = 0$. So $f_v(\phi) \in \{0, \frac{1}{2}\}$. If $f_v(\phi) = \frac{1}{2}$, also $f_v(\psi) = \frac{1}{2}$ but then in every case by definition 3 $f_v(\phi \wedge (\phi \vee \psi)) = \frac{1}{2}$. It follows that $f_v(\phi) = 0$ in which case $f_v(\phi \wedge (\phi \vee \psi)) = 0 = f_v(\phi)$.

In case (c) the thesis is trivially satisfied.

In case (d) $f_v(\phi \wedge (\phi \vee \psi)) = 0$ may be satisfied only if $f_v(\phi) = 0$.

vii.   Clearly, $\perp \in \mathcal{A}$ and for every $\psi \in \mathcal{S}$, $\perp \cup |\psi| = |\psi|$.

viii.   Clearly, $\top \in \mathcal{A}$ and for every $\psi \in \mathcal{S}$, $\top \cap |\psi| = |\psi|$.
☐

14 Distributive and modular properties

Distributive and modular properties are not satisfied by $\mathcal{A}$. To show this it suffices to recall the celebrated property by which a lattice is modular if and only if does not contain a "pentagon", that is a sublattice having the following structure (cf. Grätzer 2003: 79):

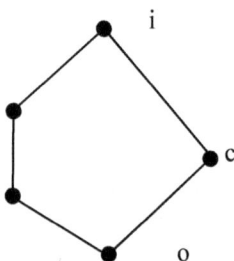

Clearly, $\mathcal{A}$ contains such a sublattice. For, given two elements b, c of $\mathcal{A}$ to which distinct atomic sentences of $\mathcal{L}$ belong, the following lattice is clearly a sublattice of $\mathcal{A}$:

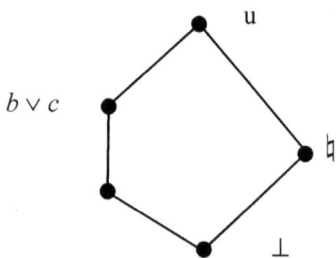

We may ask whether $\mathcal{A}$ is a semidistributive lattice. Recall that a lattice is said to be *semidistributive* if and only if the two following implications are satisfied (Grätzer 2003: 353)

(SD$_\wedge$)   if $u = (x \wedge y) = (x \wedge z)$ then $u = (x \wedge (y \vee z))$
(SD$_\vee$)   if $u = (x \vee y) = (x \vee z)$ then $u = (x \vee (y \wedge z))$

Unfortunately, these implications are both violated in $\mathcal{A}$. For, suppose $x = \natural$, $y = |q|$, and $z = \neg y$ for an atomic sentences q. Then:

$$\bot = x \wedge y = x \wedge z = \natural \wedge y = \natural \wedge z \neq x \wedge (y \vee z) = \natural \wedge (y \vee \neg y) = \natural \wedge \top = \natural$$

$$\top = x \vee y = x \vee z = \natural \vee y = \natural \vee z \neq x \vee (y \wedge z) = \natural \vee (y \wedge \neg y) = \natural \vee \bot = \natural$$

The failure of distributive and semidistributive properties may appear very unpleasant. However, we have to sacrifice some of the "nice" algebraic properties of the original de Finetti's algebra $\mathscr{L}$, if we are aiming to achieve a structure which is not isomorphic to $\mathscr{L}$. Moreover, since both $\mathscr{L}$ and $\mathcal{A}$ are structures that generalise Boolean algebras, and a Boolean algebra is just a bounded distributive complemented lattice, we have no choice if we aim at minimizing our departure from Boolean algebras: either sacrifice distributivity or sacrifice complementation. De Finetti's original approach sacrifices complementation retaining distributivity. We shall see that our approach while sacrificing distributivity retains complementation (which however, can be no longer unique).

15 Complementation and orthocomplementation

It is easy to verify that very element of $\mathcal{A}$ has a complement. This is proved by the following result:

THEOREM 11. For every element $x \in \mathcal{A}$ there exists element $y \in \mathcal{A}$ such that (a) $x \vee y = \top$ and (b) $x \wedge y = \bot$.

Proof. If $x \neq \natural$, $x \vee \neg x = \top$ and $x \wedge \neg x = \bot$ (as it may verified either directly by definition 3 or by mutant truth-tables). If $x = \natural$, then for every factual y, by Theorem 7, $x \vee y = \top$ and, by Theorem 8, $x \wedge y = \bot$. $\square$

Theorems 7 and 8 show that $\natural$ is an *almost universal complement* in $\mathcal{A}$ according to the following definition:

Definition 10. Given a bounded lattice L, an element u of L is said to be an *almost universal complement* iff for every element $x \in L - \{u, \top, \bot\}$ it holds: $u \vee x = \top$ and $u \wedge x = \bot$.

It is also easy to verify that $\natural$ is the only almost universal complement of $\mathcal{A}$. Moreover, $\natural$ is self-dual, in the sense that $\natural = \neg \natural$.

By departing from the distributive property in favour of complementation, our semantics takes the same direction as quantum algebraic logic, whose most important properties are *orthocomplementation* and *orthomodularity*. The problem arises whether $\mathcal{A}$ is orthocomplemented.

Definition 11. A lattice L is said to be an *involution lattice* if it is equipped with a map $g : L \longrightarrow L$ which is an *involution*, i.e. it holds for every $x \in L$, $x = g(g(x))$.

Clearly, $\neg$ provides such an involution for $\mathcal{A}$.

Definition 12. The involution g of an involution lattice L is said to be an *order inverting involution* iff for every $x, y \in L$ such that $x \leq y$, $g(y) \leq g(x)$.

Again, it is straightforward to verify that $\neg$ provides such an order inverting involution for $\mathcal{A}$. We are now ready to define orthocomplementation.

Definition 13. A bounded lattice L with an order inverting involution $g : L \longrightarrow L$ is said to be *orthocomplemented* iff for every $x \in L$, $g(x)$ is a complement of x.

Now, $\mathcal{A}$ is *not* orthocomplemented, due to the presence of the self-dual element ♮. It seems quite natural to provide a more general notion by restricting the condition of the preceding definition to those elements of L that are not self-dual. I shall call the resulting lattices *almost orthocomplemented*, according to the following definition.

Definition 14. A bounded lattice L equipped with an order reversing involution $g : L \longrightarrow L$ is said to be *almost orthocomplemented* iff for every $x \in L$ such that $x \neq g(x)$, $g(x)$ is a complement of x.

It is straightforward to verify that $\mathcal{A}$ is an almost orthocomplemented lattice.

16 De Morgan and Kleene's properties

It is well known that every orthocomplemented lattice equipped with an order inverting involution $g : L \longrightarrow L$ satisfies the so-called "De Morgan rules" (see Rédei 1998: 35), namely:

For every $x, y \in L$, (a) $g(x \vee y) = g(x) \wedge g(y)$, and (b) $g(x \wedge y) = g(x) \vee g(y)$. We shall show that this is true also for almost orthocomplemented lattices since De Morgan rules involve only the presence of an order inverting involution.

THEOREM 12. Let L be a lattice equipped with the order-inverting involution $\neg$. Then (a) $\neg(x \vee y) = \neg x \wedge \neg y$ and (b) $\neg(x \wedge y) = \neg x \vee \neg y$

Proof. Let $x, y \in L$. Then:

(a) $x \leq x \vee y$ and $y \leq x \vee y$. Since $\neg$ is order inverting $\neg(x \vee y) \leq \neg x$ and $\neg(x \vee y) \leq \neg y$, so that $\neg(x \vee y) \leq \neg x \wedge \neg y$. Now, $\neg x \wedge \neg y \leq \neg x$ and $\neg x \wedge \neg y \leq \neg y$, so that $x \leq \neg(\neg x \wedge \neg y)$ and $y \leq \neg(\neg x \wedge \neg y)$. It follows that $x \vee y \leq \neg(\neg x \wedge \neg y)$, so that $\neg x \wedge \neg y \leq \neg(x \vee y)$.

(b) $x \wedge y \leq x$ and $x \wedge y \leq y$. Since $\neg$ is order inverting $\neg x \leq \neg(x \wedge y)$ and $\neg y \leq \neg(x \wedge y)$, so that $\neg x \vee \neg y \leq \neg(x \wedge y)$. Now, $\neg x \leq \neg x \vee \neg y$ and $\neg y \leq \neg x \vee \neg y$, so that $\neg(\neg x \vee \neg y) \leq x$ and $\neg(\neg x \vee \neg y) \leq y$. It follows that $\neg(\neg x \vee \neg y) \leq x \wedge y$, so that $\neg(x \wedge y) \leq \neg x \vee \neg y$. $\square$

Theorem 12 shows that in $\mathcal{A}$ De Morgan laws are preserved. Another important property of de Finetti's original algebra is the so-called Kleene's condition, according to which for every x, $x \wedge \neg x \leq x \vee \neg x$. The following result shows that Kleene's condition is preserved in $\mathcal{A}$.

THEOREM 13. For every $x \in \mathcal{A}$ it holds: $x \wedge \neg x \leq x \vee \neg x$.

Proof. Since $x \wedge \neg x$ is a quasi-contradiction, either $x \wedge \neg x = \natural$ or $x \wedge \neg x = \bot$. Since $x \vee \neg x$ is a quasi-tautology either $x \vee \neg x = \natural$ or $x \wedge \neg x = \top$. In every event $x \wedge \neg x \leq x \vee \neg x$. $\square$

17 Orthomodularity

Departing from complementation or distributivity calls for minimizing such a departure if possible. We have seen that $\mathcal{A}$ is an almost orthocomplemented lattice. We have also seen that neither modularity nor semidistributivity are satisfied by $\mathcal{A}$. Orthomodularity is a weaker form of modularity defined on an orthocomplemented lattice. A perfectly analogue condition may be defined for almost orthocomplemented lattices as follows:

Definition 15. An almost orthocomplemented lattice $\langle L; \wedge, \vee, \neg \rangle$ is called *almost orthomodular* iff the following condition is satisfied for every $x, y, z \in L$:

if $x \leq y$ and $\neg x \leq z$ then $x \vee (y \wedge z) = (x \vee y) \wedge (x \vee z)$.

We may ask whether $\mathcal{A}$ is almost orthomodular. The answer is no. To show this it suffices to pick a hypervaluation that assigns the truth-value t to every sentence in y and the truth-value u to every sentence that be-longs either to x or to z. In such a case $x \lor z = \top$, so that the truth-value of any sentence belonging to $x \lor (y \land z)$ would be u while the truth-value of any sentence that belongs to $x \lor z$ would be t.

18 Distributivity and logical independence

We have seen that none of the common attempts to preserve either dis-tributivity or modularity on $\mathcal{A}$ has been successful. In spite of this there is another condition that, when satisfied ensures the satisfaction of the dis-tributive formula. This condition is *complete independence*. Let us begin by defining complete independence.

Definition 16. A subset U of an almost orthocomplemented lattice $\langle L; \land, \lor, \neg \rangle$ is *completely independent* iff for arbitrary disjoint finite sub-sets $\{u_1, \ldots, u_n\}$ and $\{v_1, \ldots, v_m\}$ of U, $u_1 \land \ldots \land u_n \land \neg v_1 \land \ldots \land \neg v_m > \bot$.

In terms of the notion of complete independence it is possible to de-fine a weaker form of distributivity that, as we shall prove, is satisfied in $\mathcal{A}$.

Definition 17. An almost orthocomplemented lattice $\langle L; \land, \lor, \neg \rangle$ is *freely distributive* iff for any completely independent set U of L, if x, y, z are distinct elements of U then it holds:

(a) $x \lor (y \land z) = (x \lor y) \land (x \lor z)$ and (b) $x \land (y \lor z) = (x \land y) \lor (x \land z)$.

THEOREM 14. $\mathcal{A}$ is freely distributive.

Proof.[3] (a) In developing the truth-table for $x \lor (y \land z)$ taking x, y, z as primitives, no quasi-tautology may be encountered, so that the developed truth-table is identical to the original truth-table developed with de Fi-netti's truth-tables. Same story for $(x \lor y) \land (x \lor z)$. But the original truth-table of $x \lor (y \land z)$ is the same as the truth-table of $(x \lor y) \land (x \lor z)$. Delet-ing from both tables the rows for the inconsistent assignments of truth-values to x, y, z, the two truth tables remain identical. This shows $x \lor (y \land z) = (x \lor y) \land (x \lor z)$. (b) Analogous (*mutatis mutandis*). $\square$

---

[3] This proof is just sketched. Details are left to the reader.

19 Probability defined on $\mathcal{A}$

In Section 9 we have set five axioms for the probability of trievents. There the axioms were meant to be applied to de Finetti's lattice $\mathcal{L}$. However, they are satisfiable also in $\mathcal{A}$. There is a remarkable result proved in 1991 by Goodman, Nguyen and Walker (see Goodman and Nguyen 1995: 261) by which given two elements of $\mathcal{L}$ x,y for every probability function $\mathbf{P}$ (partially) defined over $\mathcal{L}$ it holds that $x \leq y$ iff $\mathbf{P}(x) \leq \mathbf{P}(y)$, provided $\mathbf{P}(\updownarrow x) > 0$ and $\mathbf{P}(\updownarrow y) > 0$.

Unfortunately, the stricter relation $x < y$ iff $\mathbf{P}(x) < \mathbf{P}(y)$ for every probability function $\mathbf{P}$ does not hold. However, the only exceptions to the satisfaction of this relation arise when x,y are either quasi-tautologies (that receive all probability 1) or quasi-contradictions (that receive all probability 0).

As we have seen, quasi-tautologies and quasi-contradictions play a crucial role in $\mathcal{L}$ and lead to paradoxes. So, for example, since $(p \mid p) \wedge (q \mid \neg p)$ is a quasi-contradiction, $\mathbf{P}((p \mid p) \wedge (q \mid \neg p)) = 0$ although $\mathbf{P}(p \mid p) = 1$ and $\mathbf{P}(q \mid \neg p)$ may be assigned any value in the closed interval $[0, 1]$. This paradox was mentioned by Edgington (2006) to show that the trievent approach to conditionals is untenable. Fortunately, the difficulty disappears if probability is distributed over $\mathcal{A}$. For, in $\mathcal{A}$ it holds that $(p \mid p) \wedge (q \mid \neg p) = (q \mid \neg p)$, so that $\mathbf{P}((p \mid p) \wedge (q \mid \neg p)) = \mathbf{P}(q \mid \neg p)$ in accordance with common intuition.

Since quasi-tautologies and quasi-contradictions in $\mathcal{A}$ are (if they are not nullities) in the same equivalence class as tautologies and contradictions, the relation $x < y$ iff $\mathbf{P}(x) < \mathbf{P}(y)$ holds as is proven by the following results.

THEOREM 15. Let $\mathbf{P}_A$ be a probability function partially defined over $\mathcal{A}$. Given two elements x,y of $\mathcal{A}$ such that (a) $\mathbf{P}_A(\updownarrow x) > 0$ (b) $\mathbf{P}_A(\updownarrow y) > 0$, (c) $x \leq y$, it holds that $\mathbf{P}_A(x) \leq \mathbf{P}_A(y)$.

Proof. Let $f : \mathcal{L} \longrightarrow \mathcal{L}$ be a surjective homomorphism such that for every $x \in \mathcal{L}$ $f(x) = \begin{cases} \top & \text{if } \natural < x \\ \bot & \text{if } x < \natural \, . \\ x & \text{otherwise} \end{cases}$

It is immediately verified that $f(\mathcal{L})$ is isomorphic to $\mathcal{A}$, so we may identify $f(\mathcal{L})$ and $\mathcal{A}$ ($\mathcal{A}$ is indeed a sublattice of $\mathcal{L}$). Moreover, there is a bijection between the probability functions over $\mathcal{L}$ and the probability functions over $\mathcal{A}$. Since if $\natural < x \leq \top$ then for each probability function $\mathbf{P}_{\mathcal{L}}$ over $\mathcal{L}$ it holds that $\mathbf{P}_{\mathcal{L}}(x) = \mathbf{P}_A(f(x)) = 1$. Analogously if $\bot \leq x < \natural$ then for each probability function $\mathbf{P}_{\mathcal{L}}$ over $\mathcal{L}$ it holds that $\mathbf{P}_{\mathcal{L}}(x) = \mathbf{P}_A(f(x)) = 0$.

For each probability function $\mathbf{P}_{\mathcal{L}}$ over $\mathcal{L}$ corresponds a probability function $\mathbf{P}_A$ over $\mathcal{A}$ such that for every $x \in \mathcal{L}$ it holds that $\mathbf{P}_{\mathcal{L}}(x) = \mathbf{P}_A(f(x))$. On the other hand it holds that $x \leq y$ iff $f(x) \leq f(y)$. Now, we know that x and $y \in \mathcal{L}$, $\mathbf{P}_{\mathcal{L}}(\updownarrow x) > 0$ and $\mathbf{P}_{\mathcal{L}}(\updownarrow y) > 0$, then $\mathbf{P}_{\mathcal{L}}(x) \leq \mathbf{P}_{\mathcal{L}}(y)$ for every probability function over $\mathcal{L}$ (see Goodman and Nguyen, 1995: 261). But being $\mathbf{P}_A(f(x)) = \mathbf{P}_{\mathcal{L}}(x)$ it follows that if x and $y \in \mathcal{L}$, $\mathbf{P}_{\mathcal{L}}(\updownarrow x) > 0$ and $\mathbf{P}_{\mathcal{L}}(\updownarrow y) > 0$, then $\mathbf{P}_A(\updownarrow x) > 0$ and $\mathbf{P}_A(\updownarrow y) > 0$. Moreover, if for any probability function $\mathbf{P}_{\mathcal{L}}(x) \leq \mathbf{P}_{\mathcal{L}}(y)$ then $\mathbf{P}_A(x) \leq \mathbf{P}_A(y)$. $\square$

THEOREM 16. Given two elements x, y of $\mathcal{A}$, if for every probability function $\mathbf{P}$ partially defined over $\mathcal{A}$ it holds that $\mathbf{P}(x) = \mathbf{P}(y)$, then $x = y$.

Proof. If for every probability function $\mathbf{P}$, $\mathbf{P}(x) = \mathbf{P}(y) = 1$ then $\mathbf{P}(\uparrow x) = \mathbf{P}(\updownarrow x)$ and $\mathbf{P}(\uparrow y) = \mathbf{P}(\updownarrow y)$. Since $\uparrow x$, $\updownarrow x$, $\uparrow y$, and $\updownarrow y$ belong to the Boolean algebra $\mathcal{D} = \{x \in \mathcal{A} : x = \uparrow x\}$, it holds that $\uparrow x = \updownarrow x$ and $\uparrow y = \updownarrow y$. Now, $\{x \in \mathcal{A} : \uparrow x = \updownarrow x\} = \{\top\}$, so that $x = y = \top$. If for every probability function $\mathbf{P}(x) = \mathbf{P}(y) = 0$, $\mathbf{P}(\neg \uparrow x) = \mathbf{P}(\updownarrow x)$ and $\mathbf{P}(\neg \uparrow y) = \mathbf{P}(\updownarrow y)$. But $\{x \in \mathcal{A} : \uparrow x = \neg \updownarrow x\} = \{\bot\}$, so that $x = y = \bot$. Otherwise, supposing for some probability function $\mathbf{P}(x) = \mathbf{P}(y)$ and $x \neq y$, then either $\uparrow x \neq \uparrow y$ or $\updownarrow x \neq \updownarrow y$. There are two cases (i) $\uparrow x \neq \uparrow y$, (ii) $\uparrow x = \uparrow y$ while $\updownarrow x \neq \updownarrow y$. In the case (i) there is a probability function by which $\mathbf{P}(\uparrow x) = 0$ and $\mathbf{P}(\uparrow y) \neq 0$, so that $\mathbf{P}(x) = 0$ and $\mathbf{P}(y) \neq 0$. In the case (ii) $\mathbf{P}(\uparrow x) = \mathbf{P}(\uparrow y)$, but for at least a probability function $\mathbf{P}(\updownarrow x) \neq \mathbf{P}(\updownarrow y)$, so that $\frac{\mathbf{P}(\uparrow x)}{\mathbf{P}(\updownarrow x)} \neq \frac{\mathbf{P}(\uparrow y)}{\mathbf{P}(\updownarrow y)}$ and $\mathbf{P}(x) \neq \mathbf{P}(y)$. In both cases we have a contradiction with the assumption $\mathbf{P}(x) = \mathbf{P}(y)$. $\square$

Theorem 15 shows that probability defined over $\mathcal{A}$ is fully monotonic with respect to the lattice natural order. Theorem 16 proves that if two elements x, y of $\mathcal{A}$ receive the same probability value by every probability function that assigns to x, y both a value, then $x = y$. Taken together these results show that probability, defined over $\mathcal{A}$, preserves the lattice order in the same way as classical probability does in Boolean algebras. This fully vindicates the idea that probability over trievents is a generalization of the logic of trievents and that conditional probability may be considered as the probability of de Finetti's conditional, albeit with a new semantics, which is truth-functional in a generalised sense.

References

Adams, E. W. (1975). *The Logic of Conditionals. An Application of Probability to Deductive Logic.* Dordrecht-Boston: Reidel.

Belnap, N. D. Jr. (1973). "Restricted Quantification and Conditional Assertion". In *Truth, Syntax and Modality*, ed. by H. Leblanc. Amsterdam: North-Holland, pp. 48-75.

Bergmann, M. (2008). *An Introduction to Many-Valued and Fuzzy Logic.* Cambridge (UK): Cambridge University Press.

Blamey, S. ([1986] 2002). "Partial Logic". In *Handbook of Philosophical Logic*, ed. by D. Gabbay and F. Guenther, 2nd edition, vol. 5. Dordrecht-Boston: Kluwer.

Bochvar, D. A. ([1937] 1981). "On a Three-Valued Calculus and Its Applications to the Analysis of the Paradoxes of the Classical Extended Functional Calculus". *History and Philosophy of Logic* 2, pp. 87-112. Translated by M. Bergmann from "Om odnom Tréhznačnom Isčislénii I égo Priménénii k Analizu Paradoksov Klassičéskogo Rasširénnogo Funkcional'nogo Isčisléniá". *Matématčéskij Sbornik* 4 (46), pp. 287-308.

Calabrese, P. G. (1994). "A Theory of Conditional Information with Applications". *IEEE Transactions on System, Man, and Cybernetics* 24 (12), pp. 1676-1684.

Cleave, J. P. (1991). *A study of Logics.* Oxford: Oxford University Press.

de Finetti, B. ([1936] 1995). "The Logic of Probability". *Philosophical Studies* 77, pp. 181-90. Translated by R. B. Angell from "La logique de la probabilité", *Actualités Scientifiques et Industrielles* 391, Paris: Hermann et Cle, pp. 31-9. *Actes du Congrès International de Philosophie Scientifique*, Sorbonne, Paris, 1935, IV. Induction et Probabilité.

de Finetti, B. (2006). *L'invenzione della verità.* Milan: Cortina.

de Finetti, B. (2008). *Philosophical Lectures on Probability*, collected, edited and annotated by A. Mura. Springer.

Edgington, D. (2006). "Conditionals". *Stanford Encyclopedia of Philosophy*, ed. by E. Zalta. URL: http://plato.stanford.edu/entries/conditionals/.

Goodman, I. R. and Nguyen, H. T. (1995). "Mathematical Foundations of Conditionals and their Probabilistic Assignments". *International Journal of Uncertainty, Fuzziness and Knowledge-Based Systems* 3 (3), pp. 247-339.

Grätzer, G. (2003). *General Lattice Theory.* 2nd edition, Basel: Birkhäuser.

Jeffrey, R. C. (2004). *Subjective Probability: The Real Thing.* Cambridge (UK): Cambridge University Press.

Lewis, D. K. (1976). "Probabilities of Conditionals and Conditional Probabilities". *Philosophical Review* LXXXV, pp. 581-9.

Łukasiewicz, J. ([1930] 1967). "Philosophical Remarks on Many-Valued Systems of Propositional Logic". In *Polish Logic: 1920-1939*, ed. by S. McCall. New York: Oxford University Press, 1967, pp. 66-87. Translated by H. Weber from "Philosophische Bemerkungen zu mehrwertigen Systemen des Aussagenkalküls". *Comptes rendus des séances de la Société des Sciences et des Lettres de Varsovie* 23, cl. iii, pp. 51-77.

Malinowski, G. (1993). *Many-Valued Logics*. Oxford: Oxford University Press.

Milne, P. (1997). "Bruno de Finetti and the Logic of Conditional Events". *British Journal for the Philosophy of Science* 48, pp. 195-232.

Milne, P. (2004). "Algebras of Intervals and a Logic of Conditional Assertions". *Journal of Philosophical Logic* 33, pp. 497-548.

Rédei, M. (1998). *Quantum Logic in Algebraic Approach*. Dordrecht-Boston: Kluwer.

Reichenbach, H. ([1935] 1949). *The Theory of Probability*. Berkeley: University of California Press. Translated from *Wahrscheinlichkeitslehre* by E. H. Hutten and M. Reichenbach, Leiden: Sijthoff.

van Fraassen, B. C. (1966). "Singular Terms, Truth-Value Gaps, and Free Logic". *Journal of Philosophy* 68, pp. 481-95.

Appendix on

Bruno de Finetti, *L'invenzione della verità*

**(written in 1934, first published
by Cortina, Milan, 2006)**

The Story behind the Book

FULVIA DE FINETTI

To the Royal Academy of Italy, Rome

The undersigned Dr. Bruno de Finetti, lecturer (*libero docente*) in Infinitesimal Calculus at the R. University of Trieste, begs to enter his study entitled

THE INVENTION OF TRUTH

a typescript of which is enclosed, for the scholarship competition (*premi d'incoraggiamento*) on the fifth subject set by the said Royal Academy (studies on the "Forms and dimensions of experience" (space, time) viewed from a triple standpoint — psychological, physical and gnoseological).

Most respectfully,

Trieste, 16th November 1934/XIII

The year-book for the Royal Academy of Italy VII-VIII-IX 1934-1937-XIII-VX contains the Regulations set by the National Education Minister for awarding such scholarships. They extend to 13 articles, art. 7 of which states that, after a summary inspection of entries to decide which should be disqualified on grounds of lateness or other formal impediment, the Academic Board shall assign acceptable entries, as well as any pro-

Bruno de Finetti, Radical Probabilist
Maria Carla Galavotti (ed.)
Copyright © 2009

posals from academicians, to the various categories of the Academy to which they may belong.

Unlike the *Accademia dei Lincei*, which has to this day been divided into two categories or *classi*, the Academy of Italy contemplated four: moral and historical Science; Letters; Arts; physical, mathematical and natural Science. My research at the archives of the *Accademia Nazionale dei Lincei*, which houses the Historical Archive of the Royal Academy of Italy Collection, revealed that *The invention of truth* is still filed under Attachments to the category of Letters. One assumes from this that the Academic Board examining the entry assigned it to that *classe*, no doubt going more by the title than by its author's position as a lecturer in Infinitesimal Calculus. The study was hence examined by *literati* rather than philosophers or scientists who would have been a fairer jury had it been classed by its content as belonging to Moral and historical Science or Physical, mathematical and natural Science. The Academic Board's task can hardly have been easy, of course, given the avalanche of entries (1579) which were thus assigned: 518 to *Scienze morali*, 243 to *Scienze fisiche, matematiche e naturali*, 590 to *Lettere*, 228 to *Arti*. For the record, 54 scholarships went to Moral Science, 11 to the category of Physical, mathematical and natural Science, 23 to Letters and 16 to Arts. A glance down the *Lettere* prize-winners will fail to find the name of Bruno de Finetti.

Though it is unremarkable in the circumstances that the study should not have won a "premio d'incoraggiamento", it is somewhat surprising that my father did not try and get it published at some later date, despite keeping a flimsy of the typescript on which he made some trifling amendments. Maybe some of his diffidence about publishing was due to memories of the initial difficulty he had had some years before in publishing *Probabilism*, a mathematico-philosophical work which is fundamental to his thought and of unquestioned importance in the history of science. I well remember how, as I laboured years ago to send off my father's archive to the Hillman Library of Pittsburgh University, who had asked me for it to add to the "Archives of Scientific Philosophy in the Twentieth Century", alongside those of Rudolf Carnap, Hans Reichenbach and Frank P. Ramsey, I came across a letter signed fr. Agostino Gemelli O.F.M., rector of the Catholic University of the Sacred Heart, dated 30th May 1930 and addressed to Dr. Bruno de Finetti, Central Institute of Statistics, via Santa Susanna 17, Rome. In that missive Gemelli told my father he was willing to consider publishing *Probabilism*, though reserving judgment until he had examined the manuscript which should be sent forthwith. On 17th July 1930 the manuscript was returned with a polite covering note from the Statistics Laboratory of the Università Cattolica. Though the work had merits, "it did not appear suited to the character of the Reviews that the Catholic University edits and especially not to the Review of Neo-Scholastic Philosophy whose policy and traditions can but make it recoil from any subjectivistic approach to the problem of knowledge". It concludes: "I am sure you will have no difficulty in placing your study with some mathematical or also philosophical journal". *Probabilism* was published a year later in 1931 by *Logos Biblioteca di Filosofia*, edited by Antonio Aliotta.[1]

A thin thread of "ostracism" or benevolent neglect links *Probabilism* to *The invention of truth*. Their common fate was no doubt due to a section of the then dominant culture being averse to such daring innovations finding their way into scientific thought.

My father appended the following précis to his study: "After proving the illusory and harmful nature of philosophical concepts other than empiricism, and briefly summarising the requisite tools of formal and probabilistic logic as viewed from the chosen standpoint, the critique addresses the formation and value of key points in any scientific conception of the world, especially the notions of time and space".

Some of the views expressed in *The invention of truth* are to be found in others of his works, but to my mind those so far published contain no such extensive or specific philosophical argument. I therefore feel it deserves to be made public where it may inject new life into the current debate.

---

[1] For a deeper review of the events leading up to publication of *"Probabilismo"* see the recent article by Luca Nicotra, "Bruno de Finetti scrive a Adriano Tilgher", in *Lettera Matematica Pristem* 64, July 2007.

On Bruno de Finetti's
L'invenzione della verità

DONALD GILLIES

1   Importance of the work

De Finetti's essay: *L'invenzione della verità* is of very great importance both as a philosophical work in its own right, and also for understanding the philosophical background to de Finetti's theory of probability. It is therefore very surprising that it has only been published 72 years after it was written. In almost all his works on probability (even some of the most mathematical), de Finetti includes some philosophical observations. For him philosophizing was an important part of doing mathematics. However, this is the only work in which de Finetti presents his overall philosophy of science and mathematics. His overall position is very interesting in itself, and contains, as we shall see, many novel features. However, it is also interesting because it gives the background to de Finetti's views on probability. Indeed he himself describes the essay (p. 94)[1] as:

> showing also (as was perhaps necessary to make my point of view on this subject understandable) how my opinions on the foundations of the theory of probability can be framed as particular consequences of a more general vision of the meaning of science and of logic.

---

[1] Page references with no further specification are to the Cortina edition of de Finetti's essay. The English translations from the Italian are my own, but have been checked by my wife: Grazia Ietto Gillies.

Bruno de Finetti, Radical Probabilist
Maria Carla Galavotti (ed.)
Copyright © 2009

So the framework for de Finetti's subjective theory of probability is now available for the first time.

2   De Finetti's general philosophy of science and mathematics

The year 1934 when de Finetti's essay was written, was certainly one of great significance for the philosophy of science. The Vienna Circle had been in existence for 12 years and in 1929 had published its manifesto. 2 years after 1934, Schlick's assassination would lead to the rapid dispersal of the Circle. 1934 was also the year of the publication of Popper's *Logic of Scientific Discovery*, a work which both criticized and developed the Vienna Circle position. De Finetti also developed the Vienna Circle's position, but, as we shall see, in a direction opposite to that of Popper. While Popper stressed objectivity and attempted to rehabilitate metaphysics, de Finetti moved towards subjectivism and a more radical empiricism. The influence of the Vienna Circle on de Finetti's essay is clearly shown in the citations he makes. On page 123, he refers to Carnap's 1928 *Der logische Aufbau der Welt*, and on page 143 to Reichenbach's writings on relativity. However there are citations of some philosophers of science outside the Vienna Circle. On page 138, he refers to Bridgman's work in the German translation of 1932 (*Die Logik der heutigen Physik*).[2] Indeed operationalism played an important part in the development of de Finetti's views on probability, where, in the early version of the theory at least, degrees of belief are given an operational meaning through betting. De Finetti also cites Poincaré on several occasions (e.g. 138).

Let us now survey some of the main features of de Finetti's radical empiricism.

2.1   Links between science and philosophy

De Finetti thinks that there should be strong links between science and philosophy. He writes (71) of:

---

[2] It is interesting to note that de Finetti in the 1930s could read French and German easily, but was not so familiar with English. In his 1938 article, he discusses Keynes's *Treatise on Probability*, and says (363, Footnote 18): "I briefly saw Keynes's book in 1929 ... understanding little of it, however, because of my then insufficient knowledge of English. This year I have read the German version."

Philosophy which ... detaches itself from science and from the problems which scientific progress continually raises, thus depriving itself of the only possible source of nourishment, and condemning itself to sterility and fossilization in the monotonous repetition of sentences which become ever more empty.

This brief quotation gives a good idea of the charming and polemical style in which de Finetti's essay is written. It also reinforces the contemporary relevance of the essay, since, certainly, much philosophy today is completely detached from science and scientific problems.

2.2   *Metaphysics is meaningless*

De Finetti follows the Vienna Circle in holding that metaphysics is meaningless and accepting the verifiability criterion of meaning. As an instance of the latter, he claims on page 134 that:

> ... in speaking of the probability that a law will hold for ever, we are uttering a sentence without meaning: we can ask if the law will still hold for a certain number, even immense, of years, but the passage to the limit is illicit.

We see here de Finetti reacting to a problem which had arisen with the Vienna Circle's criterion of verifiability, because universal laws are not verifiable. Popper, who had no doubt that universal laws were both meaningful and necessary for science, concluded that the verifiability criterion of meaning was inadequate. De Finetti, by contrast, continued to accept the verifiability criterion and concluded that universal laws had to be replaced by more limited claims about the future. This is a move towards a more radical empiricism.

As a strong empiricist, de Finetti is strongly critical of the Kantian *a priori*, and polemicises against Kant in the following passage (75):

> And no one should be surprised if also Kant, who was not even a geometer, was not able to imagine non-Euclidean geometries, as no other person was able to do so before Bolyai and Lobachevsky, neither if he was incapable of reaching concepts different from the only ones then known regarding time and causality. But he was wrong to mistake the poverty of his imagination for the fount of 'a priori' certainty, in the same way as the country bumpkin who rules out the sphericity of the Earth because he cannot imagine the

inhabitants of the antipodes suspended from the Earth with their heads down.

The relevance of this today is again obvious, since the attempt to revive the Kantian position, in a new form, is quite a popular research programme in contemporary philosophy of science.

2.3   Logic and the world built up from sensations

Another feature of de Finetti's radical empiricism is that he attempts to build up logic (including probability) and the world using sensations alone from a Cartesian-type premise, namely (100): "I have sensations". Of course this fits very well with de Finetti's subjectivism, but, in carrying out his programme, he does have to modify this subjectivism at one point by allowing the sensations of other people as well as his own. Thus he writes (124) that he is:

admitting ... that the primary material which I have at my disposal is not constituted only by my experience, but also by that of others of whom I have information.

In line with this approach based on sensations, de Finetti identifies degree of belief with a sensation, saying (115):

the sensation ... will consist in feeling a certain degree of belief in the occurrence of the given event.

And again (117):

the evaluation of a probability depends on a feeling.

However this makes him liable to Ramsey's objection that (1926: 169):

This view ... seems to me observably false, for the beliefs which we hold most strongly are often accompanied by practically no feeling at all; no one feels strongly about things he takes for granted.

3   Novel features of de Finetti's empiricism

3.1   De Finetti's treatment of mathematics

De Finetti agrees with the Vienna Circle's logicist view that mathematics can be reduced to logic. He praises mathematical logic as (82) an "ingenious and important creation in which Giuseppe Peano played such a large part." He goes on to say (83):

> For a being of infinite intelligence the most beautiful and difficult of our theorems would be as banal and empty as is, for us, the affirmation that '1/2 = 0.5", where, having introduced two ways of referring to the same thing, one says that they refer to the same thing. Mathematics is true only because it does not say anything substantial, because it is purely tautological; ... This does not mean despising or diminishing mathematics: on the contrary, it is the most perfect, fertile, and powerful instrument which our intelligence commands, ...

So far, de Finetti goes along with the Vienna Circle, but now he introduces a new idea. He maintains that, although mathematics consists of definitions and conventions, it does contain something empirical, though not in the sense of John Stuart Mill. Thus he says (83-84):

> Identical truths are only formal truths, resulting from definitions and conventions freely made by us, and in only one sense can one look for something substantial in them: that is one can look for how and why we are convinced of the desirability of introducing such conventions and definitions. This examination carries us back to empirical ground, it carries us therefore to conceive also of logic and mathematics empirically. Not in the sense of falling again into the error of Stuart Mill, that is considering logical and mathematical truths as experimental truths, but, however, distancing ourselves decisively also from the most restricted form of rationalism – the mathematical rationalism, particularly characteristic of Russell – for whom at least mathematical truths would constitute a realm of pure pertinence of the intellect."

We see here the appearance of some pragmatic attitudes regarding logic and mathematics. This kind of pragmatism plays an important role in de Finetti's overall philosophical position, as we shall see.

Since de Finetti rejects universal laws as unverifiable and hence meaningless, we would expect him to have doubts about the infinite in mathematics. This is indeed the case, as is shown by his discussion of the infinite in Section 16, 112-13. However, de Finetti's doubts do not lead him to adopt the position of strict finitism as regards mathematics. The infinite is indeed so useful for the mathematician that a brilliant mathematician like de Finetti would be reluctant to give it up. This is what he says about the matter (112-13):

> Should we therefore refuse any extension of logic and mathematics beyond the restricted limits of finitism? Perhaps not. One can think of the attempt as the study of logic for a hypothetical mind where the order of succession of logical processes could be transfinite; up to what point and in what sense such an aspiration can be justified for minds necessarily finite like ours is a delicate question to examine and discuss, but a rejection *a limine* [from the threshold – D.G.] does not seem justifiable.

3.2 De Finetti's novel notion of philosophical analysis and his view that concepts have a temporary character

De Finetti frequently stresses that our existing concepts are satisfactory only because of the way the world is, and that, if there were changes in the world, our existing concepts would become useless and be abandoned. He illustrates this using the concept of time, writing (126):

> And so if all the clocks and pendulums and revolutions of the stars would not maintain any stable relation between their movements, each one would give us a different measure of time, … If there came to be lacking … that agreement between many procedures for measuring the same magnitude, which precisely makes it seem useful to us to conceive and define as a unique magnitude the result of those measurements, the invention of the concept would not have occurred. I could always eliminate every doubt rendering the definition unambiguous, for example by measuring time with a single preselected clock or with the Earth's rotation; however on the day

on which the clock or the Earth began to turn with a rhythm altogether capricious compared to that of the other presently regular movements, the definition would not suffer, being a definition, but all usefulness would vanish from it.

The Vienna Circle, and many subsequent philosophers, held that the main activity of philosophy should be the logical analysis of concepts. De Finetti does not altogether disagree, but, because he thinks that the usefulness of concepts depends on the empirical situation, he argues that the usual concept of logical analysis should be replaced by something richer. As he says (127):

> The logical analysis which I hold to be most useful is therefore that which dissects a concept by deepening the examination of how and why it could have seemed useful to us to invent it, the examination of the hypotheses on which it depends and of the possibilities of their refutation.

De Finetti's concept of logical analysis has not been widely adopted in philosophy of science, but this seems to me a pity. The subject would benefit if more practitioners used this approach to logical analysis rather than the more usual one.

Since concepts depend on the empirical situation, they are liable to change with the advance of science. De Finetti mentions the change form Euclidean to non-Euclidean geometry and Einstein's introduction of a new concept of simultaneity as examples of this process. This emphasis on the temporary nature of concepts in science leads him to introduce an analogy which is quite similar to one employed by Popper in the very same year. Popper writes in his (1934: 111):

> Science does not rest upon solid bedrock. The bold structure of its theories rises, as it were, above a swamp. It is like a building erected on piles. The piles are driven down from above into the swamp, but not down to any natural or 'given' base; and if we stop driving the piles deeper, it is not because we have reached firm ground. We simply stop when we are satisfied that the piles are firm enough to carry the structure, at least for the time being.

De Finetti in his (1938: 146) writes:

One sees that all is built on moving sand, although naturally one seeks to place the supports on the points which are relatively less dangerous ...

Despite the great differences between the 1934 philosophical systems of Popper and De Finetti, this seems to be one point on which they are in agreement.

3.3   De Finetti's view of logic and the world as 'inventions'

Popper always held to a strongly realistic view of science, but de Finetti disagrees with him on this. For de Finetti realism is an illusion and both logic and the world are 'inventions', though not arbitrary inventions. As he puts it (92):

Regarding the word *invention*, it seems to me the most appropriate for avoiding the possibility of thinking of interpretations based on the illusion of realism; it does not mean, however, that truth, logic and the world may be invented on a whim, in an arbitrary fashion, but only that in the elaboration of such concepts thought has an active part, and is guided by considerations of utility, not constrained by necessary laws.

At one point here de Finetti is in agreement with Kant against whom he polemicises on other occasions. Kant would certainly agree that in the elaboration of our concept of the world, thought has an active part. However, Kant would also hold that this elaboration is constrained by necessary laws, and not merely guided by considerations of utility.

3.4   De Finetti's anticipation of Wittgenstein's view
that meaning is use

Finally it is interesting to note that de Finetti at one point anticipates the view of meaning as use which was later to be developed by Wittgenstein in his *Philosophical Investigations* (1953). In fact de Finetti writes (105):

... for the significance of any concept the "operationalist point of view" (*punto di vista operativo*) is valid in a certain sense, that is to say that it is valid to ask the question "in what cases do I make use of it?", "why is it useful to me to make use of it?" ...

Later on, de Finetti explicitly attributes the "operationalist point of view" (*punto di vista operativo*) to Bridgman, writing (138):

... the problem is well illuminated in the book (*Die Logik der heutigen Physik*. German translation, München, 1928) by an eminent American physicist, P.W. Bridgman, in which one can see more clearly and with notable fullness the more strictly physical and contemporary side of the questions connected with the operationalist point of view (*punto di vista operativo*).

References

Bridgman, P.W. (1927). *Die Logik der heutigen Physik*. German translation of *The Logic of Modern Physics*, München, 1932.

Carnap, R. (1928). *Der logische Aufbau der Welt*, Berlin: Weltkreis Verlag.

de Finetti, B. (1938). "Cambridge Probability Theorists". English translation in *The Manchester School of Economic and Social Studies* 53, 1985, pp. 348-63.

Popper, K.R. (1934). *The Logic of Scientific Discovery*. 6th revised impression of the 1959 English translation, Hutchinson, 1972.

Ramsey, F.P. (1926). "Truth and Probability". In *The Foundations of Mathematics and other Logical Essays*, ed. by R.B. Braithwaite. Routledge and Kegan Paul, 1965, pp. 156-98.

Wittgenstein, L. (1953). *Philosophical Investigations*. English translation by G.E.M. Anscombe, Oxford: Blackwell, 1963.

Truth and *Certainty* in Bruno de Finetti

GIORDANO BRUNO

Thanks to Bruno de Finetti we don't have to invent the "Truth" any more! Readers of *L'invenzione della verità* will probably agree with me, since this book frees us of the need to have an absolute, unassailable "truth".

I have experienced this personally, and far from distressing it was. On the contrary: I felt my mind infinitely opened to the possibility of amazing new knowledge, and myself transported into a beautiful "Cantorian" garden.

Why was I not daunted? Why was I so delighted? A famous book by Morris Kline came to mind: "The loss of certainty", which is about what we usually call the crisis of the foundations of mathematics. But de Finetti's interpretive key was there to reassure me: I would not be thrown into the primordial chaos, I would not lose the path of knowledge, our sole guide to the certainties we need to live and plan our lives by.

The absolute "truth" had been replaced by a new "certainty", weaker but maybe more vivid and creative. In Bruno de Finetti's words:

> As a boy I began to grasp that the idea of 'truth' is incomprehensible ... So – case by case, more or less unconsciously – I began analysing what we essentially mean when we use the common locution that "something is true". Only now is my thirst to dissect this problem properly slaked. Besides mathematical logic (especially the theory of nominal definition) and the positive critique of the empirical world – where I found much that conformed to my own ideas and hence was of great help in developing them – I have recently added a third and definitive mainstay to my viewpoint:

Bruno de Finetti, Radical Probabilist
Maria Carla Galavotti (ed.)
Copyright © 2009

probabilism. This corrects and integrates the other two on points that I was unable to accept: points where something or other apparently had to be seen as endowed with absolute value, transcending the psychological value it had for me, and regardless of it. (de Finetti 2006: 69)

Nowadays the advent of statistical mechanics, quantum theory, wave mechanics has come to question causality and determinism, shattering the magnificent isolation of scientific prediction and drawing it by gradual concessions closer to the common predictions or conjectures of practical life. Scientific prediction no longer holds absolute value: there is only a certain probability which at most may grow big enough to deserve the name of practical certainty. (de Finetti 1989: 77)

This "certainty" permits us to go about our business, and we should let it guide us, aware though we must be that it is only probability – albeit "a very high probability" – and therefore might fail sometimes, but only to let some other certainty blossom and make us see things from a different point of view.

Here is an example of what we may mean by the word "certainty" – an example taken not from the scientific but the literary world – and the two are ultimately not so different, as you will see. Let me show you Leonardo Sciascia's method of explaining the disappearance of Ettore Majorana.

Sciascia was up against what we generally call a "mystery" – "mystery, which is the only reality", as de Finetti used to say, quoting Bontempelli about Pirandello. I must say that de Finetti too was a great esteemer of Pirandello. For example, in 1937 de Finetti published an article "Luigi Pirandello maestro di logica" in the magazine *Quadrivio* and later in the Trento newspaper *Il Brennero*. Again, his article "Tre personaggi della Matematica: i numeri *e*, *i*, π" published in *Le Scienze*, clearly draws on Pirandello's *Sei personaggi in cerca d'autore*.

We can also draw a parallel between de Finetti and Majorana: the two were born in 1906 and started to study engineering in 1923 – de Finetti in Milan, Majorana in Rome – at the youthful age of 17. They would both decide to give up engineering, even though at an advanced stage, in order to devote themselves to their own real interests: mathematics and physics respectively.

To return to Majorana's death. Everybody agreed it was a "mystery", something that belonged to the dimension of the uncertain – and it is only by doing so, that is to say bringing the mystery into the realm of the uncertain, that Majorana's disappearance may be explained.

Sciascia proceeded in this way: first, he made assumptions about the disappearance, giving three possible explanations – suicide, abduction, voluntary disappearance.

From the very first pages of Sciascia's book *La scomparsa di Majorana*, we see how the police – and especially Senator Bocchini – were immediately convinced that Majorana had committed suicide. Bocchini refused to consider any other alternatives. Indeed, he thought that different conclusions – though possible – were highly improbable, the outcome of dangerous madness. Sciascia writes:

> Bocchini, who had had time to acquaint himself with the case, had probably reached the conclusion his experience and training would dictate: here as ever two kinds of madness were combined – that of the missing person and that of his family. Science, like poetry, is known to be on the very threshold of madness. And the young professor had overstepped that threshold and jumped into the sea or Vesuvius or chosen some even more fanciful way of killing himself. And his family, as always happens when the corpse isn't recovered, or is only recovered later, by chance, unrecognizable, is seized with the madness of believing he's still alive. And that madness would gradually subside if it wasn't constantly fed by madmen who turn up and say they've seen the missing person, identified him by unmistakable signs...

On examining the note sent to the police-headquarters of Naples and Palermo, he points out that:

> ... the whole communication seems to imply: note that it's the family that requires renewed investigations; we are convinced that the professor committed suicide – God only knows how and where. And just as no result was reached in the first place, so no result will emerge from renewed investigations.

The police are thus led by the "reasoning of certainty": the suicide of Ettore Majorana is a fact, no other assumption deserves to be considered. In any case:

And even if Majorana hadn't killed himself but was in hiding, then the problem consisted in finding a madman. Thus there was little point in mobilizing men to discover a corpse – that could only turn up by chance – or a madman who, sooner or later would be seen and reported anyhow (according, as ever, to experience and statistics[1]).

Sciascia realizes a fundamental intuition here: we cannot draw conclusions about what actually happens by simply basing ourselves on statistics. De Finetti had the same intuition when he founded probabilistic inference on Bayes's theorem. According to this well-known theorem, there is a proportion between the final probability of a hypothesis explaining a fact and the initial probability of the hypothesis, multiplied by the probability of the same fact conditioned by that hypothesis ("that is to say the probability that can be considered on a statistical basis").

This is what one should do when investigating a murder. In this case the situation is very delicate, because – first of all – you must decide how to divide up the hypotheses carefully, without immediately rejecting the more unlikely ones; in fact what looks improbable at first sight might later become very likely, depending on changes in the set of information gathered.

This is exactly the method followed by Sciascia: first he considers all possible hypotheses as to the disappearance of Majorana and expresses his own opinion on each of them; then he concentrates on peculiar facts and events in Ettore's life, trying to fathom his nature and feelings; he finally comes to formulate what he thinks may be the most probable hypothesis – "Majorana had decided to escape from the world to find his own peace in a monastery"!

Sciascia chose the following passage from his book – Majorana himself is speaking – in order to illustrate his own thought in "The value of statistical rules in physics and social sciences" (*Valore delle leggi statistiche nella fisica e nelle scienze sociali*):

A radioactive atom's disintegration can force an automatic reactor to register it with a mechanical effect made possible by adequate

---

[1] Or rather: once more the experience, once more the statistics (my own English translation of Sciascia claim: "ancora l'esperienza, ancora la statistica", which I think it is much more explicative for the sake of my reasoning).

amplification. Thus ordinary laboratory equipment is sufficient to prepare an extremely complex and showy sequence of phenomena "set off" by the accidental disintegration of a single radioactive atom. From a strictly scientific point of view there is nothing to stop us from considering it as plausible that an equally simple, invisible and unpredictable vital phenomenon might be the cause of human existence.

It is a clever reflection upon chaotic effects – effects that are not linear – and the importance of events that are "quite improbable". According to de Finetti, even if an event has only a small probability of coming true, it does not mean that it will not come true. "Our degree of belief" (probability) is just a measure of the possibility.

Finally, let us read Sciascia's own conclusion on the disappearance of Majorana:

In all things, on the contrary, there is a "rational" mystery of essences and correspondences, a tight, uninterrupted network of almost imperceptible, almost inexpressible significances linking one point to another, one thing to another, one being to another. As Nisticò repeated for my benefit the unexpected, unsuspected, incredible news the distant voice of his friend had imparted to him, I experienced a revelation, a metaphysical, mystical experience. I was rationally convinced, beyond and above rationality that, whether or not they corresponded to the true, verifiable reality, those two insubstantial facts converging on a single spot could not be without significance. Nisticò's suspicion that the 'famous scientist' mentioned thirty years ago by Brother Misasi might have been Majorana; the rumour that an American officer overcome by remorse at having commanded or been a member of the crew of that fatal plane, had come to that same monastery and was perhaps still one of its inmates – how could one not link these two facts, see them as a reflection one of the other, as explaining each other, as constituting a revelation?

This "rational conviction" is the selfsame "practical certainty" that de Finetti refers to: the result of a hard search for something to be accepted as a "surrogate for the truth" (if you like this expression), for a truth that maybe does not even exist or has to be "invented", because the only principle that leads us to accept or reject it is its "usefulness".

In other words, we have to decide whether or not it can fulfil our own needs, and therefore we have to be totally aware of the "responsibility" we bear when making a choice.

References

de Finetti, B. (1989). *La logica dell'incerto*, ed. by M. Mondadori. Milan: Il Saggiatore.

de Finetti, B. (2006). *L'invenzione della verità*. Milan: Cortina.

Sciascia, L. (1975). *La scomparsa di Majorana*. Turin: Einaudi. English translation: Rabinovitch, S. (2004). *The Moro Affair and the Mystery of Majorana*. New York Review Book.

Name Index

A

Acocella, N. 153, 158, 166, 184
Adams, E.W. 201, 241
Adriani, R. 141, 151, 159, 184
Afriat, S.N. 150, 151
Albini, G. 2
Aliotta, A. 247
Allais, M. 115
Amaldi, U. 3, 5
Amoroso, L. 2
Angell, R.B. 241
Armendt, B. 67, 81
Arrighi, C. 21
Arrow, K. XIII, 115-117, 122, 124-126
Atkinson, A.B. 170, 171, 184

B

Bayes, T. 50, 62, 81, 85, 192, 262
Belnap, N.D. Jr. 213, 241
Bergmann, M. 218, 241
Berkeley, G. 20, 21
Bernays, P. 4
Bernoulli, J. 23, 24, 27, 38, 62, 116, 117
Bernstein, A. 15
Berzolari, L. 5

Birkhoff, G. 2, 5, 68, 75, 83
Blair, T. 171
Blamey, S. 203-205, 207, 210, 216, 241
Bocchini, A. 261
Bochvar, D.A. 207, 208, 241
Bodiou, G. 75, 76, 80, 81
Bohm, D. 26
Bohr, H. 5
Bohr, N. 4
Bolyai, J. 251
Bompiani, E. 3, 5
Bontempelli, M. 260
Borel, É. 2, 15, 50
Born, M. 4
Bortolotti, E. 2, 5, 6
Brandolini, A. 170-173, 184
Bravais, A. 196, 199
Bridgman, P.W. 21, 38, 250, 257
Brier, G.W. 101, 103, 104, 106, 107, 110, 112
Brodie, B. 7
Brown, G. 171
Bruno, G. IX, XIV, 259
Bub, J. 81
Burgatti, P. 5